Working with Concrete

Working with Concrete

RICK ARNOLD

The Taunton Press

The Taunton Press
Inspiration for hands-on living®

The Taunton Press, Inc., 63 South Main Street, PO Box 5506, Newtown, CT 06470-5506
e-mail: tp@taunton.com
Distributed by Publishers Group West

Editor: David Schiff/Nailhaus Publications, Inc.
Cover design: Cathy Cassidy
Interior design and layout: Jeff Potter/Potter Publishing Studio
Illustrator: Ron Carboni
Photographers: Rick Arnold, with the exception of: ©Dan Rockhill, p. 87 (right); Roe A. Osborn, courtesy *Fine Homebuilding*, © The Taunton Press, Inc., pp. 108, 109, 110, 111, 112, 114, 115 (left); © Bruce Greenlaw, p. 131 (top); © Matt Larratt, p. 168.

Library of Congress Cataloging-in-Publication Data
Arnold, Rick, 1955-
    Working with concrete / Rick Arnold.
        p. cm.
    ISBN 1-56158-614-5
    1. Concrete construction. I. Title.
    TA681 .A73 2003
624.1'834--dc21

Printed in the United States of America
10 9 8 7 6 5 4 3 2 1

The following manufacturers/names appearing in *Working with Concrete* are trademarks:
Big Foot®, Dig Safe ®, Dow®, Masonite®, Styrofoam®, Tyvek®.

About Your Safety: Homebuilding is inherently dangerous. From accidents with power tools or hand tools to falls from ladders, scaffolds, and roofs, builders and homeowners risk serious injury and even death. We try to promote safe work practices throughout this book, but what is safe for one builder or homeowner under certain circumstances may not be safe for you under different circumstances. So don't try anything you learn about here (or elsewhere) unless you're certain that it's safe for you. If something doesn't feel right, don't do it. Look for another way. Please keep safety foremost in your mind whenever you're working.

To the sons of R. Arnold & Sons Concrete Forms, Inc., Rob and Nate, who were excited about being included in the business name until that first summer vacation they spent carrying forms

ACKNOWLEDGMENTS

A lot of people put up with me constantly calling, harassing, and being a real nuisance on their job sites in order to get the photos for this book. These guys are real craftsmen here in Rhode Island: Frank and Anthony Nappa of A&F Concrete Forms, Inc.; Mark Nelson of Nelson Brothers Construction; Jeff Picard of JP Construction; Paul Corpus of Yankee Flatworks; Shawn Ward, Paul Forsell, and Matt Mirandou of Cedarhurst Construction, Inc.; David Peterson and Mike Keegan of D&M Concrete Forms, Inc.; Chic, Peter, and Mark Gervasio of A. Gervasio Construction, Inc.; Scott Grenon of Grenco Construction; Bob Chew of RemodelWrights; and Rob Arnold of New England Construction Services. The local business owners I annoyed were John Courtois of Heritage Concrete and Dave Courtois of Adamsdale Concrete. Thanks to Dick Pastore of RP Engineering, Inc. for his magic with the CAD drawings. The crew at The Taunton Press, including Peter Chapman and Carolyn Mandarano, did their usual excellent job, and special thanks to editor David Schiff, who came riding in on his white horse to save the day.

Contents

Introduction

Have you ever watched a foundation crew work? They strain and sweat, lugging endless stacks of heavy, oily forms off a large flatbed truck down into the hole. They get covered with the release oil and concrete dust left on the forms from the previous job as they set them into position. In the summer, there is no shade and no relief. If the wind does make it down into the hole, it carries clouds of teeth-gritting dirt. In the winter, clothes go through freeze/thaw cycles as the work alternates between the heavy carrying and setting to the less-physical measuring, squaring, and checking.

The pressure is constant because the concrete trucks have been scheduled to arrive at a certain time and everything must be 100 percent complete. One missed piece of hardware can cause a catastrophe.

The next morning after the pour, the crew returns to strip the job and lug all the forms up out of the hole and back onto the flatbed, only to drive to the next job and start all over again. Early in my construction career as I watched a foundation crew, I said to myself, "If there is one job I never want to do . . ." Oh, if only I had a crystal ball back then.

In the mid 1990s, the foundation subcontractor I had been using decided to move on. He offered me a good deal on his trucks and equipment and he even included a couple of excellent workers. So I decided to supplement my construction business with a foundation company. The foundation business quickly dominated, and for the last six years, 80 percent of my work has been pouring hundreds of foundations.

So now I wear three hats—general contractor, framer, and foundation contractor. It occurred to me that my experience in these roles gives me just the perspective I needed to write this book.

As a general contractor or construction manager, I address the design of the foundation as it relates to the characteristics of the building lot. A lot of money can be wasted trying to radically alter the contour of the surrounding land to accept a poorly thought-out design. Also, I explain the strategies to prevent either groundwater or surface runoff from becoming a problem. The cost to cure a leaky basement is astronomical compared with the effort required to prevent the problem. And, of course, there are always the contractor's budgeting questions. Why so much for angles? How much to add steel? Why should I pay for stronger concrete? Etc. . . .

As a framer, I'm concerned with the details that will make life easier. A square and level foundation really helps to get things off to a smooth start. Correctly sized door openings, beam pockets, and anchor-bolt placement all contribute to reducing the aggravation a framer faces when he begins to work on top of a foundation. I also review the head scratching involved to determine which way to set a foundation out of square as is sometimes necessary when adding on to an existing out-of-square house.

As a foundation contractor, I explain all of the basics from concrete mixtures to forming and pouring. I put emphasis on a lot of common-sense practices and procedures and the reasoning behind them. I have poured foundations for many good builders and craftsmen who have very little knowledge in this area. As with any other trade in the construction process, the more you understand it, the better you will be with designing, budgeting, and scheduling.

In this book, I also cover flatwork (floors, walkways, patios, etc.) enough to give you a good understanding of how to form, pour, and finish it. The smaller walkways and patios are relatively easy. But if you plan to take on a larger project such as a garage or basement floor, be careful: It takes some experience to get the knack of concrete placement and finishing.

The few books about foundations that I have thumbed through over the years were either of the old school in that many of the methods are just too labor intensive, or the subject matter was too technical and did not always apply in the

real world of residential construction. I wrote this book to answer all the questions I have faced as a novice and as a seasoned builder. Whether this book occupies a space in your office or gets kicked around the interior of your pickup covered with coffee stains and concrete dust, if it saves you time, steers you in the right direction, or prevents a major bungle, then I've done my job.

Good luck.

CHAPTER 1

Site Work

SITE WORK CONSISTS of all the things that need to be done to the land before a foundation is built. These include evaluating whether the soil is suitable to build on, clearing the site, providing access for equipment, laying out the foundation, and excavation.

In my view, site work is the least understood of the building trades, and it is where a lot of contractors and owners go wrong, resulting in the loss of a lot of money (see sidebar, p. 21). Site work should be approached with the same care and precision you would devote to any aspect of building.

The discussion of site work in this chapter pertains to house foundations, unless otherwise noted. Of course, I can't cover all situations here, but if you work with knowledgeable subcontractors you shouldn't encounter any problem that can't be solved.

Determining Soil Suitability

You've no doubt heard that a house is only as good as the foundation it sits on. To carry that a little further, a foundation is only as good as the soil it sits on. All foundations, except those sitting directly on top of bedrock, eventually will settle to some degree. The important thing is to build on soil that can bear the weight of the house so that the settling isn't enough to cause structural damage to the foundation and the rest of the house.

A large excavator makes quick work of a foundation hole.

There are three classifications of soil based on the size of the granules or particles it's composed of. In general, the larger the particles, the better the soil is to build on. Gravel and sandy soil contain the largest particles, followed by silty soils (sometimes referred to as fines). The smallest particles are found in clay. In fact, individual clay particles are so small they cannot be seen by the naked eye.

Bearing capacity

Bearing capacity—the term used to describe how well the soil can support weight, is measured in pounds per square foot (psf). The higher the bearing capacity, the less the foundation will settle over time (see the illustration on this page).

A typical two-story house requires a bearing capacity of about 3,000 psf. This includes the weight of the building materials (dead load) and the weight of people and objects in the building (live load). Hard-packed gravel, with a bearing capacity of 4,000 psf to 8,000 psf, can easily bear the weight of this typical house. In fact, in areas where code allows, a 12-in.-wide foundation wall can be poured without a footing—the wide strip of concrete that is the foundation of the foundation (see p. 45).

Gravel or sandy soil from loose to medium-packed will have a bearing capacity ranging from 1,500 psf to 5,000 psf. Silt has a low load-bearing capacity—less than 1,500 psf. Clay usually bears 3,000 psf to 4,000 psf unless it's dry and hard, in which case the capacity increases to 8,000 psf or more. The bearing capacity of soft clay, on the other hand, will be about the same as silt, which is 1,500 psf or lower.

If the soil bearing capacity is poor, one solution is to widen the footings to further distribute the load. Another solution is to dig deep enough to hit good soil; then you can either fill with good-

Loose, sandy soil drains well but requires a larger digging area due to the pitched banks it creates.

Bearing Capacity of Soil

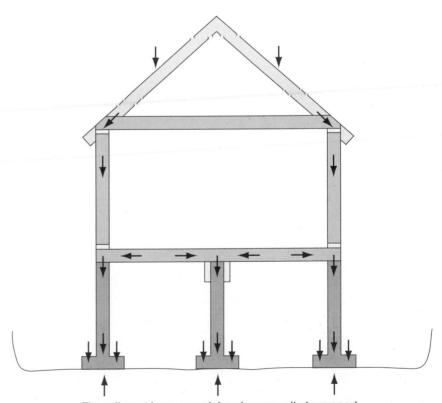

The soil must have enough bearing capacity to support the loads transferred to the foundation.

Lateral Forces of Soil Against a Foundation

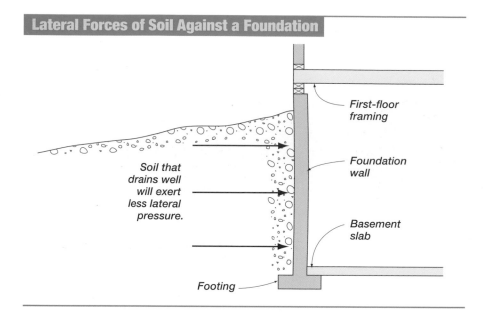

First-floor framing

Foundation wall

Basement slab

Soil that drains well will exert less lateral pressure.

Footing

bearing soil back up to the desired grade or pour taller foundation walls and backfill up to the desired cellar bottom.

Reaction to moisture and temperature

It's important to know how well the soil drains because water can significantly alter the characteristics of some soils, affecting their suitability (see the illustration above).

This lot has been partially filled to street level in the background, as evidenced by the depth of the tree. Eventually the tree will be cut and the rest of the lot brought up to grade.

Gravel and sand allow water to flow through easily, which is good for several reasons. First, it allows the gravel to compact well, which translates into increased bearing capacity. Second, good drainage means gravel and sand won't expand much when they get wet or even freeze. Finally, when rain or melted snow soaks into the ground, it increases the lateral pressure against the foundation. Well-drained soil exerts much less lateral pressure than soils that retain water.

Silt drains water poorly and, worst of all, expands when it gets wet or freezes. The lateral pressure against the foundation can be great and, as the silt dries, it can become brittle and unstable as well.

Clay not only drains poorly and retains water but also swells quite a bit in the process of rehydrating (filling up with water). One type, "fat clay," can swell so much that it can lift and cause damage to the foundation. Also, when clay dries it shrinks, which can also make the foundation move—and crack.

Other soil problems

Some soils, usually topsoil, contain dead plant life that loses its volume and settles as it decomposes. The bearing capacity of this material is very low and is not suitable as a foundation base. Here are some other problems to be aware of.

FILL MATERIAL Fill is excess soil brought in from another site. Many times vacant lots are used as disposal sites for excess material from several other sites. This means the disposal lot will have many different types of soil. If the disposal site is left alone for years after it has been filled, vegetation grows, often hiding any evidence of the fill. People often buy a lot without realizing that it had been filled years earlier.

As you excavate, you'll be able to see signs of fill in the form of thin layers of different-colored soil. Or you'll see a layer of organic material (the original ground cover and vegetation) sandwiched between two other types of soil well below the relatively new topsoil.

The best way to ensure a stable soil base for your foundation is to excavate down below the original topsoil into "virgin" ground. Depending on the amount of fill brought in, this may mean digging well below the bottom of the cellar. Then you can either fill and compact or pour tall walls.

COAL In certain geological areas, coal can be found layered between other soils. It has very little bearing capacity and should be excavated out.

PEAT Peat is found in many areas. Like coal, peat has little bearing capacity and should also be excavated out.

HIGHLY ALKALINE SOIL Soil that is highly alkaline requires you to use Type V cement in the concrete mix. You will have a very adverse effect if you use

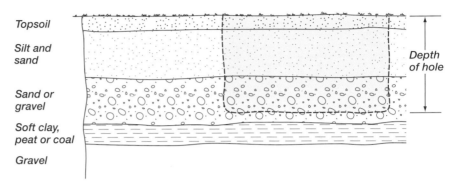

Hidden Layers of Bad Soil

Topsoil

Silt and sand

Sand or gravel

Soft clay, peat or coal

Gravel

Depth of hole

Sometimes a layer of soil that bears poorly, such as soft clay, peat, or coal, is sandwiched between layers of good soil.

Type I, which is general-purpose cement (see p. 27 for more on cement types).

HIDDEN POOR SOIL Sometimes a layer of good soil such as gravel or coarse sand covers a layer of poor soil such as soft clay, peat, or coal. When the weight of the house and foundation bears down on the gravel or sand, it in turn bears down on the poor soil, which moves, also moving the soil and house above it (see the illustration above).

To extend down below the poor topsoil, the foundation sits on top of a 2-ft.- tall footing.

Uncovering ledge can ruin your day. Extra costs are involved and the schedule is delayed until a blaster can come in to break it up enough to remove.

First the blaster drills a series of holes, then he loads them with dynamite as shown here.

Dealing with bedrock

Bedrock is solid rock, sometimes referred to as ledge. You can't get better bearing capacity than bedrock: It just doesn't move at all. If bedrock is well below grade at the correct elevation for the bottom of the cellar (assuming a full foundation) and is relatively level, then you're all set. Of course, with that kind of luck you'll probably win the lottery! Unfortunately, bedrock typically runs through the ground in random streaks and elevations. Once you unearth an outcropping of ledge, it is very much like the tip of an iceberg. The deeper you excavate, the more you will uncover (see the photo at top left).

There are a couple of ways to deal with this problem depending on the depth of the bedrock. For example, if you need a 7-ft.-deep hole for your foundation and you hit bedrock at 6 ft., you could simply stop excavating at 6 ft. and build your foundation starting at that elevation. It would mean that the foundation would stick up a foot higher out of the ground but that extra height could be hidden either by adding an extra foot of soil during the backfilling or planting taller foundation plantings.

However, if you hit bedrock at 3 ft., you certainly wouldn't want an extra 4 ft. of foundation sticking up out of the ground. At this point, unless you're willing to give up your full foundation and settle for a crawl space, you have to have the bedrock blasted out. This adds a lot of time and expense to the project. Not only do you have to schedule a blaster and wait for him to get the necessary permits, but also the excavator will have to rearrange his schedule to dovetail with the blaster's. In addition, some communities require a seismologist to record the blasting impacts to make sure they stay within local limits.

One thing you don't want to do is install a foundation with part of it sitting on bedrock and the other part on soil. The part bearing on the soil will settle over time but the part on the bedrock won't, and a crack in the foundation will develop at the juncture of soil and rock.

For this reason, the most common practice is to blast and remove the bedrock a foot or more below the bottom of the desired depth of the hole. Then, using crushed stone or gravel, fill the hole back up to the desired depth, using a walk-behind vibrating compactor or a roller-type compacting vehicle to compact the new material in 8-in. to 10-in. layers called lifts as it is being filled. The result is a level, hard packed, well-draining base (see the photo below).

Dealing with a high water table

Another potential problem is a water table that's too high. The water table refers to the level of groundwater that occurs naturally below the surface of the earth. During different seasons, it gets higher or lower by several feet, typically reaching its highest at the end of winter, when rain and in some areas melting snow raise the level. This is known as the "wet season." It's important to know what the water table level is at its highest point during the wet season and then make sure the bottom of your basement hole is at least a couple of feet above it.

If the foundation hole is excavated during the dry season while the water table is low, you could look down into the hole and assume that the nice dry soil means no future water problems.

A layer of gravel is added and compacted with a machine once the ledge is blasted and removed.

The contractor knew what to expect during excavation when he saw the exposed bedrock on this lot. Since this outcropping didn't interfere with anything, it remained and became part of the landscaping.

The large boulders used to tier the landscape were from the site excavation. It's a good indication that nearby vacant lots will be similarly difficult to excavate.

However, the water table could easily rise 4 ft. to 5 ft. during the wet season, putting the new basement floor under a substantial amount of water.

Investigating Soil Problems

To determine if you have soil problems, you can hire a soils engineer. The engineer will have core drilling or test holes dug in areas of the lot that will be excavated, such as a few spots within the foundation area and some in the leachfield area (if there is one). A complete survey by a soils engineer will tell you what kind of soil and compaction to expect at the bottom of the hole and the level of the water table.

You can also do a little investigation on your own. One way is to talk to people who live in the area. Ask them not only about their own excavations but also try to get a little history of the surrounding land. Has it been flooded or been filled? Do any neighbors have wet basements or cracked foundations?

Another potentially good source of information is an excavator who has worked in the area. He will have a ton of valuable information, not only about the soil itself but also about bedrock and the local water table.

A little reconnoitering of surrounding lots may also indicate the presence of bedrock or other problems. If the terrain is hilly and you see a large outcropping of bedrock, you should assume it's also below grade—and in all the wrong places.

If the lot is in an established neighborhood, take a look at the landscaping around the homes. If it contains boulders used for tiered elevations, chances are this rock was removed from foundation holes. This indicates a good chance your excavation will be difficult (see the photo bottom facing page).

Another clue to a possible problem excavation is if the homes are raised well above the existing grade and then beefed up with fill. This may indicate a high water table or bedrock not too far down, so the foundation had to be installed some feet above one or the other.

Deciding whether to test

With this information in hand, you can decide whether you should hire a soils engineer. If your lot is in an existing neighborhood and you know the conditions in the surrounding lots are acceptable, then you can reasonably assume that your lot will also be okay. Still, there's no guarantee: Soils within yards of one another can be dramatically different. If neighbors have had bad experiences, if the lot is relatively isolated, or if it has a questionable history, then testing is called for.

Preparing the Site Plan

Once you decide to build on a particular site, you need a site plan, also called a plot plan, prepared by a civil engineer. The site plan is a map of the site that shows where the foundation goes, how deep it should be set, and other entities on the site. It's also referred to as the site engineering. In essence, it shows where to dig and not dig. Most towns require such a plan and it will become part of the paperwork that you submit to obtain a building permit.

Here's a list of things that might be drawn on a very complete site plan. Note that not all will necessarily be on it.

- ■ Property lines
- ■ The orientation and location of the house, called its "footprint"

To keep the basement above the water table, the foundation for this house was raised well above existing grade, then fill was brought in to raise the grade.

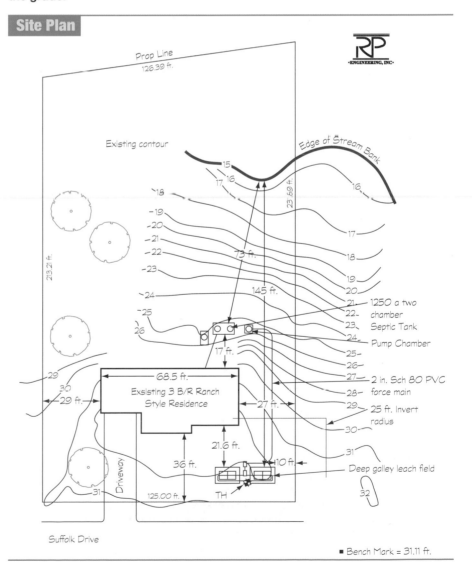

Site Plan

A gas company representative marks the location of the gas lines in front of an excavation site.

Locating Utility Lines

Before any excavation begins either on a vacant lot or an addition to an existing house, you need to locate buried utility lines to avoid damaging them. Many municipalities subscribe to a "one-call" service company such as Dig Safe®. You tell the company where you want to dig and it will contact all the utilities. The utilities send out workers to identify where their underground utilities are located. You can get the name of a one-call company from the municipality you're working in or from any excavator.

If your town doesn't subscribe to the service, you must call each utility company yourself. It's a good idea to identify the areas you will be excavating by using white paint or flags so the utilities know they have the correct site. Each type of utility is color coded for easy identification.

How long you have to wait before the all the utilities show up will vary from place to place, but the one-call service will tell you so you can include the time delay in your schedule. Also, there may be a time limit in which to complete your excavation. Sometimes it's as short as 60 days. If you have to excavate after the specified time has passed, you have to call again and repeat the process.

Color Codes for Underground Utility Lines

Each utility is assigned a different color for marking their underground lines.

▬	Electric
▬	Gas-oil-steam
▬	Communication CATV
▬	Potable water
▬	Reclaimed water
▬	Sewer
▬	Temporary survey markings
▬	Proposed excavation

- House setbacks (minimum allowable distance from the property lines)
- Height of the foundation wall
- Depth of the foundation hole
- Existing topography and proposed altered topography
- Driveway location
- Abutting road and driveway location
- Location of the sewage disposal tank and leaching field (assuming the house is not being hooked up to existing sewers)
- Water lines and hookup, if any
- Well, if there is one
- Utility lines (overhead or buried)
- Outlet pipe for the foundation drain system
- Surface drainage swales (gently sloping ditches or low areas that drain surface water)
- Curtain (also called "french") drains (trenches filled with perforated pipe and stone around the perimeter of the building to intercept and redirect surface and subsurface water)

The plan may also include what you want the excavator to remove in terms of clearing the land, but these things are often individually marked on the site with spray paint, as detailed later.

Transferring the Site Plan to the Land

The next step is to transfer the site plan to the land itself. When building a house, it's best (and in many municipalities, required) to hire a state-registered land surveyor to stake out the property. For one thing, it's a job that requires precision. For example, the municipality most likely specifies setbacks—the required distances between property lines and structures. If your plans place the building within a few inches of a setback, you don't want to risk mistakes.

Staking out

Referring to the plan as well as your oral instructions, the surveyor will use wooden stakes with orange plastic flags to mark the location on the ground of all major points. A stakeout commonly includes the corners of the property, the corners of the foundation, the septic system and leach field, and the well.

The stakes may mark the exact location of things such as the corners of the lot. But in the case of a foundation, the stakes will be offset from the exact location of a corner (see the photo below). This offset allows the excavator to dig the hole without disturbing the surveying stakes because once the hole is finished,

the exact location of the foundation will be measured from those same stakes. All stakes will be labeled, such as "Northeast lot corner." The surveyor will also establish a benchmark—a fixed point of elevation used to gauge the elevation of other items.

Using batter boards

The area within the setbacks where you are allowed to build is called the building envelope. If the building envelope isn't much bigger than the house, you need a surveyor and his transit to locate the house very accurately within the envelope. But if the envelope is a lot bigger than the house, the contractor or owner can choose to locate the house himself using batter boards and string, once the surveyor stakes out the envelope.

The surveyor's stake indicates where the foundation will be. The stake itself is located away from the excavation, so it won't be disturbed during digging.

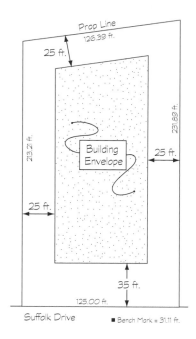

Prop Line
126.39 ft.

25 ft.

213.21 ft.

231.89 ft.

Building
Envelope

25 ft.

25 ft.

35 ft.

125.00 ft.

Suffolk Drive   ▪ Bench Mark = 31.11 ft.

A builder sets his own stakes for an addition by measuring directly off the existing house.

To set up batter boards, drive three stakes into the ground a couple of feet apart, forming a 90-degree corner near each corner of the foundation. Set the batter boards far enough back on the ground so they don't get run over by excavating equipment. Fasten horizontal boards to connect the stakes, then stretch strings across the tops of the assemblies and adjust them until an outline of the overall box of the foundation is formed (see "Laying Out Square Corners" on p. 47). Drive nails into the cross boards to mark string locations and then remove the strings so they don't interfere with the excavation (see the illustration below).

Using Batter Boards

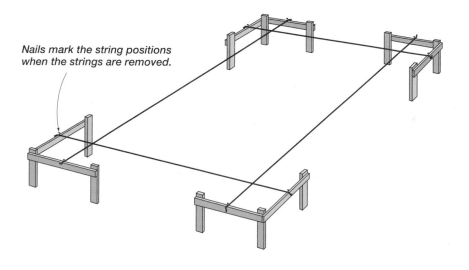

*Nails mark the string positions
when the strings are removed.*

Staking out an addition

If you are building an addition, you can easily mark out its foundation using the existing foundation as a reference. For this reason, you don't need a surveyor unless local authorities require one. The layout stakes are used as a guide for the excavator only, so they are pounded into the actual corner locations for the foundation with no offsets. After the excavation is complete, the forms are laid out by measuring off the existing foundation again (see the photo on facing page).

Clearing the Land

Before the excavating machines can start digging for the foundation, the site must be cleared of trees and large vegetation. The easiest way to determine which trees and vegetation have to be removed is to run string connecting the various surveyor stakes so you see the outline of the affected areas. The foundation, leach field, and driveway can all be outlined for easy identification.

Establish an identification system with the tree cutter so there is no question as to what is to be cut and what isn't. Remember that you also want to remove the trees outside of but near the outlined areas to create a clear buffer around your house. You wouldn't want any large trees only few feet away from your foundation. The engineering requirements for a septic system will determine the size of the cleared buffer around that area.

Also, room is needed to maneuver large trucks around the site.

Mark trees to be cut with brightly colored spray paint and tie marker ribbon around the nearby trees that are to remain. Also, set up protective boundaries around the remaining trees that are in the construction area (see the photo below).

Establishing a plan of attack

If you will be clearing the lot yourself, have a plan of attack. Clear brush, then cut the smallest trees and work your way up to the largest. This way the larger trees won't tangle in the smaller ones when they fall. It's also much safer using a

An orange safety fence establishes a boundary around a tree to help protect it during construction.

Establishing buffers and deciding which trees must go and which ones can be saved can be a delicate balancing act. The excavator and other workers want lots of elbow room, whereas the homeowner naturally wants to save trees.

One way to help decide is to hire an arborist who can identify the affected trees and describe potential problems as they relate to the new construction and practical buffers.

A better way is to hire a landscape architect to identify existing vegetation, establish buffers, and create a complete plan that encompasses existing and new plantings.

chainsaw without having to worry about tripping in brush. Stack the wood far enough away on site so it doesn't interfere with the excavator's movements. He needs room to maneuver his equipment and large areas to stack the excavated soil. Also, stack the branches in a spot that will allow you to easily burn (you may have to get a permit to do this) or chip them.

You can also hire a licensed and insured tree cutter for the job. These cutters shouldn't be confused with landscapers. You can find them in the Yellow Pages under "Tree Services." It's a good idea to walk the property with the cutter and make sure there is an understanding on the scope of the work to avoid mistakes. Many issues must be decided, such as:

- Who keeps the wood?
- Where should it be stacked?
- What lengths will it be cut into?
- Are fallen or standing dead trees outside the affected area to be included in the price?
- Do branches get stacked, burned, or chipped?
- How much time does the cutter need to do the job?
- Who clears the underbrush? (Sometimes it's better to have the excavator "scrub out" the dense brush. He rakes the ground with his machine and hauls away or buries piles of scrapings.)

Lot clearing by a tree cutter may not include getting rid of tree stumps. If not, the excavator can haul them away. Or if your municipality allows it and you have enough land, the excavator can bury the stumps on the site far away from the house. Bury them as deep as possible and leave a small mound of earth over the

The cleared wood was chipped into piles well away from the staked-out construction area on this lot.

burial site. Over time the stumps will decay and possibly collapse, sometimes dramatically, and you want the earth there to fill the depression. This possibility and the possibility of the wood attracting termites lead many homeowners to have the stumps hauled away.

Providing Access

As mentioned, clearing the site is important to provide access for the excavating equipment and concrete truck. But clearing the site is not enough. There are other things that may have to be done, particularly for the concrete truck. It's a monster—30 ft. long, 8 ft. wide, 12 ft. high and about 40 tons fully loaded. And this monster has a very poor turning radius.

Access to the site on a vacant lot is typically not a problem. Once cleared, there is nothing much left that will interfere. The biggest potential problem on a new site is hilly terrain. Steep sites present a challenge for the excavation equipment, but it's the concrete truck that is limited when it comes to climbing.

Considering the weather

Weather also has a big influence on the access to the site. It doesn't take much rain or snow to prevent a concrete truck from going up even a slight incline. In cold weather, it's best to pour first thing in the morning while the ground is completely frozen rather than in the afternoon after the sun has melted the top inch or so of mud.

If your project is an addition, there needs to be enough room for the excavation machine to maneuver. If you are not

Even with a section of fence removed, access to this backyard addition is severely limited by the shed and cedar tree.

A nice wide space between the foundation and the bank of the hole is known as the overdig. It allows room for the workers to safely install the foundation.

If excavators spent a few weeks on a foundation crew, they would realize the extra value of a properly planned and executed excavation site. But too often, you'll end up with too much overdig, too little overdig, not enough access, shallow or narrow ramps, a bottom that isn't level, or undercut sides. Any one of these will, at best, add hours to a job and, at worst, will stop the job from beginning until fixed. I try to inspect the excavation well before the foundation crew arrives so there is time to fix any problems. Showing up with the crew, trucks, and specific forms only to discover a problem with the hole can be very costly.

sure that there is, check the site over with the excavator before the machine arrives. If you're not sure whether the concrete truck can get in without a problem, contact the concrete company, which will gladly send a representative who is experienced enough to tell you definitively. The concrete company wants its trucks on the road making money, not ensnarled on your site. For more potential problems with additions, see p. 103.

Excavating the Foundation

When digging a foundation hole, the perimeter should extend farther than needed for the foundation. This extra area is called the "overdig" and it gives the foundation workers the room they need to move, stack, and set the forms (see the photo above).

The typical overdig for a full cellar is 4 ft. to 5 ft. all the way around. That's assuming that the foundation hole is no more than 5 ft. deep. If the hole is deeper than 5 ft., safety demands greater overdigs, which are detailed by Occupational Safety and Health Administration (OSHA) requirements. For holes less than 5 ft. deep, the overdig can be reduced to 3 ft. to 4 ft. Anything less than that will make working on the foundation forms slow and difficult.

Also, depending on soil type, large clods of earth may fall in from the side of a hole when the concrete trucks come rumbling up close to the edge. If the overdig is not big enough, that earth can hit the foundation forms and push them in a little. If there's even just a little concrete in the forms at that point, it will be almost impossible to slide the forms back into proper position without first opening them up and emptying out the concrete.

To mark for digging the foundation hole, measure 4 ft. to 5 ft. out from each foundation corner location and mark the spot with white spray paint. Spray continuous straight lines from one mark to the next until there is a completed outline of the foundation "footprint" plus the overdig. In other words, if the foundation is 26 ft. by 40 ft., the outline will be about 36 ft. by 50 ft.

Placing soil

Before the excavator begins, he will need to know where the excess dirt will go. If it can be used on site, then it should be placed as it gets dug or it should be temporarily piled far enough away from the hole so it doesn't get in the way.

When excavating for garages or shallow additions, inexperienced excavators may want to temporarily pile the dirt on the inside of the excavation. Don't allow this because later, after the forms are set and it's time to square the job, the dirt piles will make it impossible to run a tape measure from corner to corner to check for square (see the photo below right).

If there is no place on site to use the excess dirt, then it should be hauled away. In many cases, you can find a site nearby that needs the fill. To keep haul-

Your Life Depends on Proper Overdig

Providing the proper overdig is more than a matter of work convenience and protecting your forms. It is also important to protect your life. Working in a freshly excavated hole or trenches can be very dangerous. The OSHA standards and regulations should be followed to minimize the danger. In addition to OSHA compliance, a little common sense goes a long way.

When the hole is excavated, the exposed soil will begin to lose moisture and dry out; how quickly depends on soil type and climate. As it dries, it becomes unstable and can cave in.

Cave-ins are also a hazard when you excavate through a layer of frost. When

A narrow overdig creates an unsafe working environment as a worker squeezes in between the bank and the forms.

the afternoon sun hits the south-facing wall, the sun melts frost and, again, parts of the wall can cave in.

An overnight rain can dramatically alter the stability of the hole. Add to all the above conditions the deep vibrations of a rumbling, teeth-chattering concrete truck rolling just a few feet from the edge of the hole and you can see how easily people can get hurt.

Some of the excavated dirt is temporarily piled inside the addition area. This will make it difficult to square the forms by diagonal measuring.

ing charges down, the closer the location the better, and many times the person who needs the fill will pay the hauling charges.

Watching the water table

If there is no site engineering, the most important consideration in determining the depth of the hole is the water table.

Check with local authorities, but chances are that if they do not require any site engineering, then they probably don't care how close you put your foundation to the water table. As mentioned earlier, it's best to keep the cellar bottom at least 2 ft. above the highest wet-season water-table level.

Establishing a Benchmark

All of the existing and new elevations specified on the site plan are established by using a benchmark for reference. In most cases, the benchmark is a known elevation above sea level. For example, when the surveyor puts a nail in the tree or telephone pole with an orange flag that says "benchmark 104.5" it means that the nail is 104.5 ft. above sea level. Say, for example, the site plan determined that the existing topsoil grade is at 102 ft. and the cellar bottom should be at 95.5 ft. To find the right depth for the cellar, the excavator would set up a transit and shoot the benchmark to get a reference elevation. Then he would dig until the bottom of the hole is 9 ft. lower than the benchmark (104.5 − 95.5 = 9). Given that the existing grade is 2.5 ft. below the benchmark, the hole will be 6.5 ft. deep.

In addition to staking out the foundation, the surveyor establishes a benchmark. Here the benchmark is a nail pounded in near the base of this tree. The nail's elevation is written on the orange tape above.

Excavating shallow foundations

For a shallow foundation, you may have to excavate deeper than desired just to get below the topsoil and onto a good base. In that case, you can either fill the hole back up to the desired depth with gravel or, pour taller walls. If you use gravel, you should place and compact it in 8-in. to 10-in. lifts. In other words, place and spread 8 in. to 10 in. of gravel and then compact. Repeat the process until you're back up to the desired depth of the hole. If the water table is down far enough (at least 2 ft. below the cellar bottom), it need not be a concern and you can just pour a taller foundation, which is usually a less expensive approach than filling and compacting.

To reach a good base, the garage section of this foundation was dug down a couple of feet more than required for the garage foundation height. Fill was brought in and compacted in 10-in. lifts up to the proper level.

Estimating Excavation Cost

Without a doubt, site work is the most difficult to budget. If you're lucky, you are in an area where the conditions are consistent from lot to lot—similar water table, soil conditions, utilities, etc. If this is the case, it's relatively easy to get an accurate quote from an excavator who has worked in the area. But in many parts of the country, especially in New England where I work, conditions can drastically change from one lot to the next.

The best way to get an estimate is to meet at the site one at a time with potential excavators after the lot has been surveyed and completely marked out. Bring along an extra copy of the site plan. Also bring a list of what you want included in the bid, such as topsoil scraped and piled, stumps removed or buried, sufficient access for form and concrete trucks, french-drain material and installation, proper compaction of fill within the structure, driveway base provided, etc. List the specifications and acceptable tolerances, i.e., 5 ft. of overdig, 12 ft. wide, machine-compacted earth ramps, hole bottom level within 2 in. This way the excavators will all be bidding on the same thing, apples and apples rather apples and oranges.

If there is a significant amount of fill to be hauled in, get a per yard price instead of having the excavator try to guesstimate the amount of fill required. That way you pay for exactly what the job needs as it is needed.

If you suspect there is bedrock on the property, pay to have several test holes dug in the areas where the foundation and septic holes will be. An accurate price for removal can be established only after the depth and size of bedrock is determined. If bedrock is found to be within a few feet of the surface, don't forget to account for its removal for the underground utility trenches.

■ WORK SAFE
■ WORK SMART
■ THINKING AHEAD

In excavation work, you'll often have to alter your plans as you uncover unexpected conditions. For example, you may decide you need an extra 8 in. of crushed stone for drainage under the foundation. Just remember to dig the hole 8 in. deeper. Otherwise, the top of the foundation will be too high.

The bank of this excavation is undercut and therefore prone to caving in. When the workers start to erect the forms a few feet away, a dangerous working environment will be created.

Checking the completed hole

When completed, the sides of the foundation hole should have at least a slight pitch away from the hole. No sides should be vertical, or worse, undercut. The bottom of the hole should be within 2 in. of level, and there should be room around the top of the hole for the concrete trucks to access at least three sides of the foundation. To make life easier for the foundation workers, there should be a small, trenched out access ramp cut in from the original ground level to the bottom of the hole.

Building ramps

Contrary to what many people think, concrete does not flow easily and it does not seek its own level. If it does, it means it has been watered down way too much and will have relatively little strength. Concrete with the right proportion of

Without the dirt ramp built up by the excavator, the concrete truck's chute would not have enough pitch for the concrete to flow into the forms.

Ramps: What Can Go Wrong

If the excavator pulled out his machines before you got a chance to check the site, then you are stuck with the ramps he left for you. And unless the constant complaining by the foundation subs has had an effect on ramp design and construction, the ramps may well be too low, too narrow, too soft, and too far away.

You have two options if you want good results. Unfortunately they will both cost you. The first option is pull off the job and come back when the ramps are adequate. The second is to have a lot of workers on hand to shovel the concrete around corners during the pour.

There are two other options you should avoid because they will give you poor results. You don't want to add water to the concrete so it flows around corners and down the 40 ft. of wall you can't reach with the chutes. This will weaken the concrete. The other bad strategy is to try to shovel the concrete with your small regular crew. This will take too long because the concrete may begin to set and form weak imperfections called cold joints. It will also make you very unpopular with your crew.

■ WORK SAFE
■ WORK SMART
■ THINKING AHEAD

Like many subcontractors, the excavator is typically trying to juggle many projects at once. It is important to get a firm commitment for the start date so that the foundation can follow right behind. The closer you get to the start date the more frequently you should contact the excavator to see how his schedule is progressing. The excavator, more than any other sub, is prone to delays because of bad weather, equipment failure, or just the nature of the job: You're never sure what's hidden in the earth.

ingredients has to be worked with a shovel to get it into position.

The easiest way to get the concrete into position is to have access all the way around the hole so that the concrete truck can slowly drive along with its chute feeding concrete into the forms. This isn't always possible. Another alternative is to raise the truck high enough so you can add several lengths of chute to reach most areas with enough pitch for the concrete to flow.

Even though the trucks are tall, their reach is not very far because of the necessary pitch of the chutes. To get the truck higher, the excavator will build dirt ramps at different points around the hole. The tops of the ramps should be close to the level of the tops of the forms, and the ramps should be at least 12 ft. wide. Many excavators will build a ramp that seems wide enough to drive on, but if it isn't wide enough to spread the load, the tires will sink, leaving that 40-ton truck straddling the ramp. Of course, the ramps should also be compacted enough to withstand the weight, something the excavator can do by running his machine up and down on them as he builds them. If the ramps are too soft, the truck will plow into them rather than ride on top.

Excavating a frost wall

A frost wall is that part of a foundation that extends below the frost line specified by your local building code. If your foundation plans incorporate a "walk-out" section with a frost wall beneath it (see p. 95), the designer may actually have surveyed and shot the grades of that lot. If this is the case, the excavator can follow the plans for the location of the trench needed for the frost wall.

More often than not though, the plans are not that accurate. They were either designed without any particular lot in mind (mail-order plans, for example) or, if they were designed for a specific lot, chances are a topographic map was used in lieu of actually visiting the site. Topographic maps are not always accurate enough to use for exact placement of a foundation. In this case, you have to work with the excavator to establish the "drops" in the foundation wall. The loca-

▪ WORK SAFE
▪ WORK SMART
▪ THINKING AHEAD

Many excavators like to do a minimum amount of work up front, but this results in hours being added to the foundation job, which you may wind up paying for. They'll excavate the foundation hole and move the dirt, stumps, and rocks around just enough to get the machines and concrete trucks in. After the foundation is poured, they'll come back to backfill and clean up the site properly. This makes things difficult for the foundation crew, who will have to maneuver trucks and equipment over and around the stumps, rocks, and dirt. What should be a relatively easy job now becomes a big aggravation that takes a lot more time. Meanwhile, an extra hour or two on the excavation machine would help keep the form job on time and on budget. The list of specifications you give to the excavator for him to bid on should include a definition of sufficient access for the form and concrete trucks. Make that list part of your contract with the excavator.

The walls are "dropped" or "stepped down" so the foundation elevation matches the natural contour of the land. The walk-out section of the foundation along the back extends below the local frost line, in this case 42 in.

tion of the drops will depend on the future finish grade of the surrounding land (see the photo above).

Working with the excavator, you can size and locate the walk-out section to match the characteristics of the land contour, minimizing cost. The excavator will be able to give you estimates of any extra work or fill needed to make sure the finished grade matches the transitions in wall height (drops).

Backfilling

After the foundation is completed, the excavator will return to backfill the overdig. In a perfect world—from the excavator's point of view—the basement slab is already poured, the first-floor deck has been completed, and the foundation has cured for about a month. This means that the tops and bottoms of the foundation walls are locked in place, and pres-

sure from the added earth won't affect the walls.

In the real world, the schedule usually demands that backfilling be done as soon as possible. To ensure that no damage is done to the foundation, there are a variety of steps to take.

First, wait at least seven days before backfilling. This allows the concrete to cure to around 60 percent of its final strength. If the temperature is below 40°F you should wait 10 days.

Second, brace the foundation, which reinforces it against the initial shock of the earth being pushed against it. The best way to do this is to install the deck, which locks the entire foundation together at top, but this is not always practical because working around the hole adds a lot of extra time and effort to the beginning of the framing process. Plus, the framing has to stop for a day or two while the backfilling is going on. The alternative is to temporarily brace the walls using some of the framing lumber or bracing equipment manufactured specifically for this purpose.

Third, be careful about what is used for backfill. Don't use any earth containing clay, fines, or organic matter (such as topsoil), particularly if wet. As mentioned, these soils expand when wet. This can create a lot of hydrostatic pressure against the foundation walls, potentially causing cracks and water problems.

Coarse gravel makes good backfill. If there is none available on site, buy bank run or processed gravel from a quarry.

Also, keep boulders and frozen chunks of earth away from the foundation. They can crack a foundation if they bump up against the wall or are pushed against it as the earth is being dumped in.

Finally, don't allow the excavating machine to roll over any part of the backfilled area. This also can cause the foundation to crack.

When backfilling a garage, first determine the desired grade of the top of the future concrete floor, then subtract 4 in. (the thickness of the floor) and mark the inside corners of the foundation. Snap lines connecting the marks to create a continuous reference line that you backfill up to. Remember to include a slight pitch out of the building to the garage floor.

Bracing Poured Walls

If the foundation is to be backfilled before the basement slab is poured and the first-floor deck installed, it should be braced from the initial impact of the earth.

Footing form planks work well but they're usually needed at the next job, so 2-by framing stock is often used. I wouldn't make bracing from members smaller than 2 x 6.

Start by placing vertical members flat against the long walls about every 8 ft. For walls shorter than 12 ft., these members are usually not necessary. Tack each one in place with a masonry nail. Locate the verticals opposite each other, even if it means one side doesn't lay out exactly right.

Run horizontal brace stock along the bottom of the hole between the verticals. Butt the horizontal stock up against the bottom of the verticals. It's best to cut the horizontal pieces so that they butt tightly together and then nail a scab over the butt joints. But the reality is that since there often isn't power on the site at this point, so the boards often get overlapped as shown in the photo below. If you do this, be sure to overlap the members at least 2 ft. and join them with plenty of nails.

Place 10-ft. to 12-ft. diagonal braces between the vertical and horizontal members and nail them into place. The top of the diagonal should be about a foot down from the top of the foundation. To reinforce them in position, place scrap blocks of stock up against each end of the diagonal member and nail them into the vertical and horizontal members. Use plenty of 16d duplex (double-headed) nails.

Once the framers arrive, they will tear down the bracing, which has already done its job.

Basement walls are temporarily reinforced with 2-by stock against the initial impact of backfilling.

CHAPTER 2

Concrete Basics

MIX UP SOME sand, cement, and stone, add water, and you've got concrete. What could be simpler? As you'll see in this chapter though, there is a lot more to working with concrete than most people realize.

For small jobs such as deck piers and landings, you can mix your own concrete or even buy it in bags that just require you to add water. But for the quantities you'll need for even a small foundation, you'll want to order "ready-mix" concrete from a plant that mixes it to order—called batching. The ready-mix is loaded on a truck and delivered to the site. The concrete workers on site then take charge and direct the truck driver to discharge as they call for it.

Once the concrete leaves the truck's chute, there are many conditions that can radically affect the curing process. When those conditions are planned for and controlled, the finished product will have the strength and durability you paid for. If the requirements for proper handling and curing aren't met, the finished product will most likely be defective. Concrete replacement or repair can be very costly and labor intensive, but it can be avoided with the right precautions. This chapter will explain how to order and handle concrete and how it cures. It will also describe possible defects and their remedies.

A concrete delivery truck is loaded from the batch plant that combines the concrete ingredients to create concrete.

Concrete Composition

Concrete has four ingredients: portland cement, large aggregate, small aggregate, and water. The large aggregate can be gravel or crushed stone. The small aggregate is almost always sand. Portland cement comes in five types, each formulated for specific characteristics. By choosing the right portland cement type, the right size and type of aggregates, and adjusting the ratio of ingredients, you can create the best concrete for any job. Sometimes ingredients called admixtures are included to further adjust characteristics.

The purpose of aggregate

The job of the large aggregate is to increase the strength of the concrete while reducing the cost. Think of it as cement filler. The individual stones that make up the aggregate interlock with each other while the sand fills in the voids. The cement paste created from the cement and water is the glue that holds the rock and sand together.

The gravel or crushed stone that makes up large aggregate ranges in size from about ¼ in. to 2 in. across. The size of the large aggregate should not exceed one-fourth the thickness of flatwork or one-fifth the thickness of a wall. For example, if you are pouring a 4-in.-thick patio (flatwork), the stone should be no bigger than 1 in. Typically, wall mixes use a ¾-in. to 1-in. stone, whereas flatwork mixes use a ¼-in. stone (also known as pea stone). The smaller stone makes the concrete easier to work with and finish, which is a plus when pouring a large exposed surface area such as a basement or garage floor.

Small aggregate ranges from ¼-in. stone down to very fine sand. The purpose of the small aggregate is to fill the voids between the large aggregate, so it is

Aggregate of different sizes is kept in individual bins at the ready-mix plant.

important for the small aggregate to be small enough. Most mixes use sand no matter what the size of the large aggregate is.

Portland cement types

Portland cement is available in the following five types.

TYPE I GENERAL PURPOSE This is the most widely used type, especially for residential projects.

TYPE II This generates less heat than Type I, which is important in large-volume jobs because it sets up more slowly to give crews more time to work during hot weather. It is also moderately resistant to sulfates, ingredients found in deicing agents that can damage finished concrete.

TYPE III HIGH EARLY This sets and cures rapidly. It's used for jobs that must bear weight shortly after being poured. It is also used during very cold weather to reduce the time during which the concrete is prone to freezing.

TYPE IV This produces very little heat as it cures and is used for huge-volume projects such as dams.

TYPE V This is very sulfate-resistant and is used in areas in which the soil or groundwater is very alkaline.

Calculating How Much Concrete to Order

Concrete is sold by the cubic yard. The process for calculating yardage differs a little between flatwork and walls, but they both use the same formula: (Length in feet × width in feet × thickness in feet) ÷ 27.

Calculating Flatwork

For flatwork (for example, a 20-ft. by 30-ft. patio that's 4 in. thick), start by converting the thickness into a decimal fraction of a foot. In this case 4 in. divided by 12 equals 0.34 ft. So the calculation would be:

$$20 \text{ ft.} \times 30 \text{ ft.} = 600 \text{ sq. ft.}$$
$$600 \text{ sq. ft.} \times 0.34 = 204 \text{ cu. ft.}$$
$$204 \text{ cu. ft.} \div 27 = 7.5 \text{ cu. yd.}$$

Calculating Walls

For walls, begin by determining the total linear feet of the perimeter. If, for example, you have a 26-ft. by 42-ft. foundation, the perimeter is 136 lin. ft. Now you must reduce that total by the number of outside corners times the thickness of the wall because by using the outside measurements of the foundation, you have actually counted the corners twice. If, for example, the walls are 12 in. thick, multiply 1 ft. times four corners. Subtracting that 4 ft. gives you an adjusted total of 132 lin. ft. Let's say the foundation is 7 ft. 6 in. high. Here's the calculation:

$$132 \text{ lin. ft.} \times 7.5\text{-ft. wall height}$$
$$\times 1\text{-ft. wall thickness} = 990 \text{ cu. ft.}$$
$$990 \text{ cu. ft.} \div 27 = 36.67 \text{ cu. yd.}$$

I usually round up to the nearest half increment to be on the safe side. So for this patio, I'd order 8 cu. yd. For the walls, I'd order 37 cu. yd.

Ordering Concrete

Besides knowing what type you need when ordering from a concrete plant, the three major questions are: How much, what strength, and what size stone? If you want any other extra ingredients known as admixtures, which will be covered later, you must be very specific.

Each job has its own requirements for the strength, durability, and workability of the concrete. These requirements are based on two things: First, the specifications of the building codes, the project architect, or the project engineer. Second, any site conditions that can affect the outcome of the concrete in relation to those specifications for which it was mixed. In the sections that follow, you'll get a good idea of how to order concrete that has the characteristics required by the specifications and site conditions of your own projects.

Deciding on the strength you need

The strength of concrete is measured in pounds per square inch (psi). The more psi, the stronger the concrete will be. Most codes call for at least a 2,500-psi concrete to be used in residential construction. That is, the concrete must be able to withstand at least 2,500 pounds of compression per square inch before it breaks. To determine the strength you need, check with your local codes and your project documents (plans, specifications, etc.). If the strengths called for by each are different, use the stronger of the two. Not all house plans specify the strength of the concrete to be used, but if they do, and if they specify a stronger mix than the code calls for, the building inspector usually will expect you to follow the approved plans.

Curing Concrete

The chemical reaction between water and cement is called hydration. In order to maintain proper hydration, and therefore proper curing, the rate of evaporation and the temperature of the water in the concrete must be controlled. If the water rapidly evaporates or freezes during the first seven days of curing, the concrete will be substandard. This is why weather plays a big role in concrete work.

Curing in warm or hot weather

Hot, dry, windy conditions draw the water out of exposed concrete by evaporation. The more extreme these conditions, the quicker the water will be drawn out, making proper curing impossible. To prevent evaporation, the concrete must be kept either wet or covered. Saturating the work with a garden hose works well, but depending on the conditions, you may be out there several times a day keeping it wet. Once the process is started, the concrete must be kept consistently wet. Alternating wet and dry cycles can do more harm than good.

You can increase the intervals between soakings by covering flatwork with wet burlap that is manufactured in rolls for this purpose. To greatly increase the intervals, cover the burlap with polyethylene sheets. A bed of wet straw covered by polyethylene serves the same

A foundation is kept wet in hot weather to aid in proper curing.

Ponding in Hot Weather

If you are pouring flatwork during very hot, dry weather, you must keep the concrete wet until it cures properly. One effective method is called ponding. To do this for a garage or basement slab, block off the door openings and run a hose onto the slab until it is completely submerged. Top off the pond occasionally as the water leaks out. For a patio slab or walkway, after the slab has been poured and finished, run a bead of caulk on the top of the forms and add 1-in. or 2-in. stock to the top. Top the form with water and maintain it for a week.

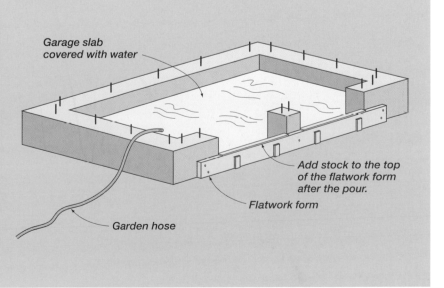

Garage slab covered with water

Add stock to the top of the flatwork form after the pour.

Flatwork form

Garden hose

A sheet of polyurethane is applied over a freshly poured patio to retain the moisture concrete needs to cure properly.

purpose, although you will have more of a mess to deal with later. White poly helps reflect the sun, thereby reducing the heat during hot temperatures, whereas black poly aids in keeping the work warm during cold weather.

Working with Curing Compound

To ensure complete coverage with a curing compound, apply it in two thin layers. Apply the second application perpendicular to the first. Use a pigmented compound to make it easy to see that you haven't missed any spots. There are pigmented compounds that disappear in a short time if left exposed.

A combination curing/sealing compound is sprayed over a freshly finished concrete walkway.

Laying polyethylene sheets directly on flatwork also keeps the water from evaporating. You need to soak the concrete first and then lay the poly flat with no air pockets. The problem with this method is that it is difficult to get rid of all the air pockets. These pockets can cause the surface to discolor. For this reason, you should reserve this method for concrete that will receive some sort of floor finish such as carpet or vinyl.

For walls, the forms should be left on for a week. Unfortunately, form subcontractors usually don't do this because they need to strip the forms the next day to use on the next job. Leaving your own site-built plywood forms on for seven days is not advisable unless you use sheathing with an overlay skin that will not bond with the concrete. That overlay sheathing is very expensive and is used for the prebuilt, panelized forms the foundation subcontractors use. Again, keeping the concrete wet will allow the concrete to cure properly. The problem is that on a vacant building lot, there is rarely a water supply nearby.

Chemical curing

In lieu of water and plastic coverings, there are many types of liquid-membrane curing compounds available that you can easily spray on any type of concrete project using a small, hand-pumped pressure sprayer. Once you apply a curing compound, moisture is sealed in and the concrete should cure properly under all but freezing weather conditions.

It's important to use the correct curing compound for the job. Depending on its base, which can be acrylic, asphalt, rubber, wax, epoxy, or some other type, curing compounds have many different side effects. For example, some will discolor the concrete, which is not good for work that is to be left exposed, such as a patio. And some leave a film that glue won't adhere to well, which is not good for glued floor finishes like vinyl sheeting or carpet.

There are also curing compounds that are supposed to seal and harden as well. Check with the supplier to get the correct compound for your job. Explain the job and the conditions so he can direct you to the best choice. Make sure that the compound will meet with the American Society for Testing and Materials (ASTM) requirements for curing, ASTM C156 and ASTM C390.

Curing in cold weather

When temperatures drop below 40°F, the curing process begins to be adversely affected. The more the temperature drops, the more critical it is to protect the concrete. Since concrete generates heat as it cures, a simple strategy of insulating and covering is sufficient. The only time you should be concerned with adding some type of auxiliary heat is when the temperature begins to fall into the single digits on the Fahrenheit scale.

Large thermal blankets are manufactured specifically for protecting concrete and are worth the investment if you will

be doing at least a few projects in cold weather conditions. A less expensive one-shot method is to cover the work with hay and black polyethylene sheeting. As with the thermal blankets, doing this will both insulate and trap moisture during the curing process.

Admixtures for tough curing conditions

If you anticipate pouring concrete under less than ideal weather conditions, discuss your concerns and requirements with the concrete plant representative. He can suggest the correct admixtures that can be added in to the concrete as it is being batched for your job.

ACCELERATORS Accelerators reduce the curing time substantially so you can shorten the length of time you need to protect the concrete from freezing. Plan to have a couple of extra helpers during the pour because the accelerator also reduces the setting time, giving you a much smaller window in which to complete the finish work.

There are also a few tricks you can use to accelerate the curing process with-

Thermal blankets are placed over freshly poured concrete during cold weather. The blankets keep the concrete from freezing by retaining the heat generated by the curing concrete.

out adding chemical accelerators. All these methods also accelerate setting. One way is to use hot water in the mix. It's common for many plants to use hot water during all cold-weather months. Another method is to use Type III High Early cement. This cures about twice as fast as the Type I cement that is normally used. Or you can increase the cement-to-water ratio. This can get expensive, so you have to weigh it against other methods.

RETARDERS   On hot days, retarders are added to increase the setting time, so there is more time to do the finish work. Retarders also help to reduce initial stress cracking that happens when concrete sets and cures too quickly. Retarders lengthen the curing time a bit, so you have to account for that in your schedule.

WATER REDUCERS   Water reducers let the concrete become more plastic and workable with less water. The benefit is a stronger concrete with less labor needed for placing and working it. High-range reducers or "superplasticizers" dramatically increase the concrete's ability to flow without reducing the strength. This is very helpful for those jobs with limited

access for concrete trucks where the concrete must be moved manually with shovels. Concrete that can flow well is less likely to develop defects such as honeycombing and cold joints (see p.39). Water reducers can be used in combination with accelerators or retarders.

AIR ENTRAINMENT   With this admixture, millions of microscopic air bubbles are created in the concrete mix. These bubbles give water in the concrete room to expand as it freezes during the freeze/thaw cycles in cold climates. It is used to make flatwork more resistant to weather and deicing agents. Air entrainment also acts as a plasticizer and allows the concrete to be more workable with less water.

Reinforcing Concrete

Any project that is cast with concrete—walls, floors, stairs, whatever—can easily be strengthened by adding reinforcement. The cost of reinforcement is minimal compared with the added value it gives. Even if reinforcement is not in the specifications, it's well worth the investment.

Structural reinforcement

When concrete specifications, local codes, or site conditions call for steel-reinforced concrete, that means adding lengths of deformed steel rod called reinforcing bar, more commonly known as rebar. Rebar is placed within concrete to control cracking and to give it more structural strength, allowing the concrete to withstand more stress and loading. Typical diameters for residential work are ⅜ in., ½ in., and ⅝ in. The ⅜ in. is more commonly used in flatwork and the ½ in.

Steel and Concrete Work Together

Take a round piece of blackboard chalk and hold it on both ends between your thumb and forefinger. Apply pressure and try to break it. You are putting the chalk under compression; like concrete, the piece of chalk has good compressive strength. Now place the chalk across two separated fingers and press down in the middle with your thumb. The chalk easily breaks. You have put the chalk under tension, and like the chalk, concrete also has poor tensile strength.

Steel, on the other hand, has excellent tensile strength, so when it is combined with concrete, the two materials create a product with great tensile and compressive strength.

Lengths of steel reinforcing rod (rebar) will be placed into the footings before the pour.

and ⅝ in. are used in walls. Proper placement and sizing will be covered in later chapters.

Crack reinforcement

Welded wire fabric (wwf) and fiber mesh are two types of crack reinforcement that have different purposes. Welded wire fabric, also called wire mesh, is used in flatwork to keep the concrete together if it cracks after it cures. It is not meant to prevent cracks but mainly to keep them from getting bigger. The 6-in. by 6-in. mesh is most commonly used and is available in 5-ft. by 10-ft. sheets or in rolls. The proper placement and techniques will also be covered in chapter 6.

Fiber mesh is mixed into concrete at the plant so the fibers are distributed through the batch. It is used to help control cracking and in many cases is incorrectly substituted for wire mesh. There are different types of fiber material from steel to synthetic, the synthetic (polypropylene among others) being more popular.

Unlike wire mesh, the fiber mesh helps prevent cracks from forming in the first place but only while the concrete is setting. After the concrete is placed, it begins to change from a plastic state to a solid. During this change, the concrete expands and contracts due to the chemical curing process and water evaporation.

Commonly referred to as wire mesh, 6-in. by 6-in. welded wire fabric (wwf) is laid in place in a patio slab.

Strands of plastic fiber are added to the concrete to help prevent cracks.

The fibers added to the concrete provide an internal network of support as the cement paste grasps them. This support reduces the amount of plastic shrinkage that occurs in concrete. Once the concrete solidifies, the fiber mesh in the concrete has done its work and is of little added benefit.

Wire and fiber mesh are good enhancements for flatwork, but they should not be used as a substitute for proper preparation, mixing, placement, and curing. Many builders think that by adding wire or fiber mesh to concrete they can cheat the requirements for a good job. A properly executed project without wire or fiber will have more strength and durability, and fewer defects, than a mismanaged pour with both wire and fiber mesh.

Testing Concrete

Testing is used to help ensure and document that the concrete used for the project meets the specifications. Although most testing occurs on commercial jobs, it is not uncommon to have testing performed on residential jobs as well. If you don't have enough experience to determine the slump by eye as it drops from the chute, you may want the assurance that testing provides.

Slump test

The slump is a measure of the consistency of concrete. A slump test indicates how much water has been used in the mix. The less water, the stiffer and stronger the mix will be.

A slump of 1 in. to 3 in.—in the trade we just say 1 to 3—indicates a very stiff mix usually called for in commercial work. This is difficult to work with and requires mechanical vibrators to get it to consolidate.

A slump of 4 or 5 is more typical in residential work. It is more workable by hand and still retains the strength that it was batched for. Using a 6 or 7 slump mix may produce a finished product that is substandard in strength and durability. Any mix with a slump more than 7 should not be used. Of course, if the mix has a plasticizer additive, then the consistency of the mix is altered without adding water, so the slump test is not an accurate gauge of strength.

Compression test

Compression tests are conducted in an off-site testing facility. A hydraulic machine applies gradual pressure to a concrete sample until it fractures. The samples are typically 6-in.-dia. cylinders, 12 in. tall. The pressure is measured in psi.

If you want to conduct compression tests for your project, purchase plastic

Doing a Slump Test

A slump test is performed on site with a sample from each truck before the beginning of the discharge. A conical metal container is filled in three equal lifts. As each lift is completed, a long ⅝-in.-dia. metal rod is placed in the mixture and pumped up and down about 25 times. The container is then gently pulled up off of and placed next to the concrete, which immediately slumps down under its own weight. The rod is placed on the top of the container extending over the slumped pile of concrete. Then a ruler is used to measure the distance between the rod and the concrete. That measurement, in inches, is the slump.

The slump test cone is filled with three equal layers of concrete. Each layer is tamped with a steel rod.

The final layer of tamped concrete is leveled off.

The cone is gently lifted off the concrete. The distance that the concrete slumps down is measured in inches from the height of the test cone.

A test cylinder is filled with concrete to be tested after it has cured.

test cylinders from the testing company and fill them up with the samples yourself. Your ready-mix supplier should be able to refer you to a testing company. The samples should be taken from each truck about mid-pour. They should also be subjected to the same curing conditions as the concrete used in the job. For example, if the conditions require thermal blankets to keep the concrete from freezing, the samples should be placed with your project beneath the blankets. Once they have cured, take the cylinders back to the testing company to have the testing performed.

If you want to test a project after the fact, you can have core drilling done, which drills out 6-in.-dia. holes in your work. The holes obviously have to be patched, but it gives you samples that can be accurately tested to determine the strength of your finished product.

Dealing with Bleed Water

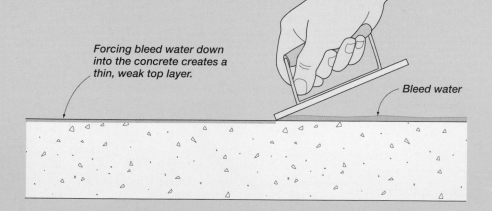

Forcing bleed water down into the concrete creates a thin, weak top layer.

Bleed water

Water that rises to the surface of the concrete after it has been placed is known as bleed water. This can be caused by working the concrete too much, such as by using a mechanical vibrator for long intervals. Bleed water can also be caused by adding an excess of water to the mix, in which case the aggregates settle and the water rises with very little encouragement. Once bleed water is on the surface, it should be allowed to evaporate off before any finishing begins. If finishing does occur before evaporation, the bleed water is forced back down, creating a weak, thin top layer that is prone to defects such as crazing and scaling.

Concrete Defects

When concrete isn't mixed, or handled, or cured properly, the result is a finished product that can have one or several different defects. Some defects appear immediately, such as surface cracks, whereas evidence of others, such as structural cracks, become apparent months or years later. By learning what they are and what causes them you can easily avoid most concrete problems.

Insufficient strength

Weak concrete is usually caused by improper curing or by mix ratios that have been altered. Following proper curing procedures will eliminate improper curing as a cause. The foundation crew can alter the mix ratios on site. The concrete truck driver has the ability to add water, stored on board, to the concrete mix before it is poured. He will only do this as directed by the person or persons pouring the job. Many foundation crews have a tendency to have the driver add too much water so they can handle the concrete more easily, but this can significantly reduce the strength of the concrete. One gallon of water added to 1 cu. yd. of 3,000-psi mix can reduce its strength by almost 200 psi.

Cracking

Cracks fall into two categories: structural and static. Structural cracks are often caused by improper soil conditions or poor control joint placement. Strategies for dealing with poor soil conditions can be found on p.4. See p.147 for information about proper joint placement. Static cracks are usually cosmetic in nature. They typically occur within the 28-day curing cycle. Static cracks can take the form of plastic shrinkage cracks or crazing.

This high-slump concrete has been watered down too much as evidenced by the way it dribbles down the chute and off the sides of the shovel.

Small plastic-shrinkage cracks were created on the surface of this stamped concrete walkway that was poured during very hot weather.

■ **WORK SAFE**
■ **WORK** SMART
■ **THINKING AHEAD**

For a good bond with any concrete repair, the surface to be worked on must be thoroughly cleaned and prepped. Use water and a stiff brush to clean the affected area. Don't worry about drying the concrete before repairing; leaving the area moist will help with bonding.

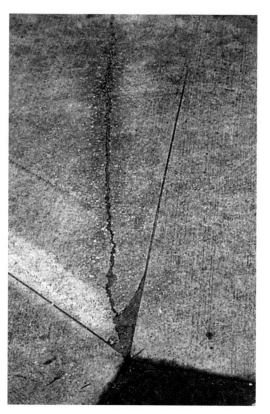

The structural crack through this slab caused by settling or frost didn't follow the control joints because the joints were too shallow.

Flatwork defects

Because flatwork is thinner than walls and has a larger proportion of exposed surface, it is more prone to certain defects. Following is a list of those defects that are more likely to occur in flatwork.

PLASTIC SHRINKAGE CRACKS These happen when the surface of the concrete prematurely dries out and hardens while the concrete beneath is still in a plastic state. As the plastic concrete moves due to expansion and contraction, the hard top surface cracks. Pouring in dry, windy conditions, especially in hot weather, evaporates the water from the surface of the concrete too quickly, causing this problem.

CRAZING This is a network of very small surface cracks usually spreading out over large areas or the entire surface. Crazing is caused by finishing the concrete with bleed water on top. The bleed water is forced back down into the surface by the finisher's trowel. This increases the ratio of water to cement, creating a weak surface layer.

SCALING Thin flakes of concrete come loose and flake or peel off the surface. The sizes of the flakes vary, but they usually increase over time and with traffic. Conditions that cause scaling include freeze/thaw cycles, deicing agents with calcium or sodium chloride, fertilizers containing nitrates, working in bleed water, or improper curing. Any one of these can cause scaling, but it's usually a combination of two or more that lead to severe scaling.

BLISTERS These are typically about ¼ in. to 1 in. in diameter, but it's not impossible to get 3-in. to 4-in. blisters. They are not easily seen until they are broken by traffic. Blisters are caused by working and finishing the surface while water or air is still working its way up through the mix to the surface.

SPALLING Spalling is similar to scaling except large chunks instead of just flakes break loose. This indicates a severe weakness in some parts or the entire project. It is more likely to happen during freeze/thaw conditions.

DUSTING Also known as chalking, this is a fine, loose powder caused by the deterioration of a weak surface. Causes of dusting include working in bleed water, improper curing, a bad sand-to-cement ratio, or exposure to carbon monoxide caused by using an unvented heat source to keep a project warm.

Wall defects

Concrete defects in walls are usually easily detectable right after the forms are removed, which is also the best time to repair the defects, before any back-filling occurs.

HONEYCOMBING This consists of areas in the concrete that are composed of large aggregate and an insufficient amount of sand and cement, which results in voids around the large aggregate. Isolated pockets of honeycombing can occur when a stiff mix (1 to 3 slump) is being used without sufficient consolidation. Or it can happen when a mix that has a high ratio of water segregates by being moved too much.

COLD JOINTS When a layer of concrete is poured and allowed to set before the next layer is added, a weak bond called a cold joint is created. A cold joint is evident by a seam containing small voids.

Voids between aggregate form a defect known as honeycombing.

Repairing Flatwork

All flatwork is subject to defects, although exterior slabs are prone to more problems than interior slabs such as basements and garages. Problems with interior slabs are usually limited to dusting and cracking. For very light surface defects such as crazing and dusting, a paint or stain with a filler can hide the problem and in most cases stop further erosion by hardening the surface. If the defects are more serious, such as scaling, spalling, and blisters, and the work will be hidden by a finished floor, you can fill the individual defects with any number of available ready-mix patching products. Either of the above fixes can be found in hardware stores, lumberyards, or large

A severe cold joint was formed when the lower section of concrete set before the next load was poured in. The strength of the wall at this point is questionable.

Preparing an Area to Be Patched

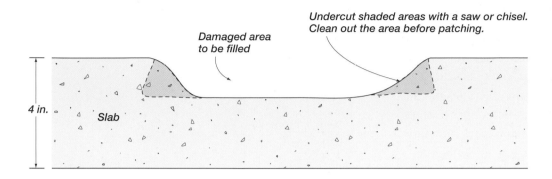

Damaged area to be filled

Undercut shaded areas with a saw or chisel. Clean out the area before patching.

4 in.

Slab

Cracks occur for many different reasons including improper mix or curing, settlement, hydrostatic pressure, stress from seismic activity, or machine damage. It is important to determine the cause of a crack before it is repaired. A crack caused by shrinkage is usually static, whereas a crack caused by settlement may continue to grow and may indicate a serious soil problem. The only way to tell is to monitor a crack for movement over time. If you have a crack that hasn't increased in size for four or five years, you may be able to repair it yourself. A contracting company that deals exclusively with concrete repair is best equipped to deal with serious, active cracks. They will have the expertise necessary to correctly diagnose a situation.

home centers. They are easy to use and work well when the directions are followed properly.

Patching

For serious defects to flatwork that will remain exposed to the weather such as driveways and walkways, there are a few remedies. You can patch the defects and then paint or stain the entire surface to hide the patches. In this case, you want to use a patch mix that has a texture that mimics the surrounding unblemished surface. Also, the affected areas should be saw-cut or chiseled out to accept at least a ¼-in.-thick patch. The sides should be undercut.

Resurfacing

Another option is to resurface the entire area. The old way of resurfacing involves pouring another 3 in. of concrete on top of the old work. If you take this approach, make sure to put some type of isolating membrane such as felt paper or polyethylene sheeting between the old and new work. This lets the new work move independently from the old. Also include control joints (see p. 147).

Since adding an extra 3 in. to a driveway or sidewalk isn't practical, resurfacing with one of the many new

resurfacing systems is the best option. Unlike patching and painting though, these systems require skill and experience to be done correctly. There are many companies and contractors that specialize in this type of covering. Depending on the system, the new top layer can be as thin as ⅛ in. and can have a compressive strength of 8,000 psi or more. As an added benefit, patterns and colors can be included in the process to produce just about any type of finished look desired.

Replacing the area

The defective area can be replaced with a new section. In this case, you should saw-cut nice clean lines completely through the slab, following the control joints. If the section is large, you must cut or break it up into manageable pieces to be removed. Then you form up the void and pour as new. This is necessary when the defects are large structural cracks caused by bad soil conditions or by external forces such as a fully loaded concrete truck driving over a patio. If the cause is a soil condition such as a loose, washed-out area, then you should fix it before the new pour.

Repairing Foundation Cracks

Of all the types of concrete crack-repair products, injection fillers are the best choice for foundation cracks. Rigid epoxy fillers bond with the existing concrete to form a union that is stronger than the concrete itself. This will prevent a crack from growing. Or a more flexible epoxy or a urethane-based filler will act as an expansion joint while retaining a good seal. The type you use depends on the diagnosis of the crack.

Regardless of the type of filler, the basic procedure, described below, is the same. There are, however, a lot of nuances in this procedure that are affected by the type of injection system you choose. The easiest and most economical systems for occasional or one-time use come as kits. They contain everything needed plus detailed instructions.

First, you must clean the crack and surrounding surface of all loose concrete, dirt, dust, or other contaminants. Water within the crack is not a problem because the epoxy will displace the water and the curing of epoxy is not affected by water. If the crack extends all the way through the wall to the outside, then any part of the crack that is above grade must also be cleaned and sealed. If poor or loose soil conditions exist below grade against the crack, the affected area may have to be excavated so you can seal the surface from the outside to contain the filler.

Then, using a small trowel, knife, or gloved hand, apply a surface-seal epoxy with a pasty consistency over the surface, bridging the crack and sealing its entire length. As the surface sealer is applied, insert small injection ports that come with the kit into the crack that will allow later injection of the epoxy filler. The

Instead of inserting injection ports directly into the crack, holes are drilled at angles on either side of the crack. Once the drill bit reaches the crack inside the wall, the drilling is stopped. Then the ports are fit into the holes.

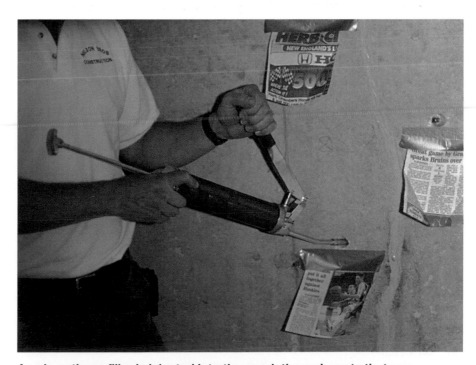

A polyurethane filler is injected into the crack through ports that were secured inside the holes. Scrap paper beneath each port protects the wall surface from any spillage.

ports should be placed at intervals that are about equal to the thickness of the wall. So if you are working on a 10-in.-thick foundation wall, the injection ports should be placed about 10 in. apart along the entire length of the crack. Build up a little extra epoxy paste around the ports to help hold them in place.

The surface seal will take 30 minutes to 24 hours to harden, depending on the type. When it is hard, inject epoxy filler into the crack, starting with the lowest injection port. When the epoxy begins to flow out of the next highest port, crimp the bottom port closed and move up to the next port and repeat the process, working your way up the crack. After the epoxy cures, the ports can be cut close to the surface and then ground flush to improve appearance.

Small static or cosmetic shrinkage cracks can be left alone unless they extend completely through a foundation wall and are either leaking water or are in a freeze/thaw climate. If surface water finds its way into a crack and freezes, a cycle begins in which the crack is widened. This, in turn, lets in more water that again freezes. Repairing the crack will stop the cycle. Excavate along the length of the crack, and fill the crack with a premixed, flexible, and watertight concrete patch available in a caulking tube. These flexible patches don't have the strength of epoxy and they won't stop a crack from lengthening.

Repairing honeycomb and cold joints

As mentioned, a honeycomb consists of a void where there should be cement paste and sand. The repair begins in the same way as all concrete repairs. Remove any loose material and thoroughly clean the area. Now you can either use regular masonry mortar or a proprietary product to fill in the honeycomb. If you use mortar, you must first wet the area and

A foundation crack is filled with caulking to prevent water migration.

brush on a thin layer of grout for a good bond. Then apply the mortar with a trowel, filling in the small voids. Some of the proprietary products are mixed with their own liquid bonding agent rather than water, or a ratio of bonding agent plus water, so you skip the first step of applying grout.

I find using a hard sponge trowel is the best tool for blending the patch into the wall. Once the patch is filled, I use a few circular movements to try to match the surrounding smooth surface.

A cold joint is filled and hidden in the same way as honeycombing. The difference between the two is that a cold joint is more apt to be a serious structural problem. Before a cold joint is filled, it should be assessed by an engineer.

Repairing an out-of-level foundation

There are instances when the transit that is used to establish the top of the foundation goes bad and the foundation is poured way out of level. Should this happen, you have a few options, all of them expensive. You can:

■ Hire a concrete cutter to cut the top of the foundation level. This means you will have to drill and epoxy in new anchor bolts in the cut areas, and the framer may have to add extra sill plates to maintain the basement headroom.

■ Hire a mason to add a custom-cut, correcting course of concrete block. This may bring the foundation farther up out of the ground than desired.

■ Form and pour a correcting layer of concrete on top of the foundation. To bond the new concrete to the old, you must drill holes into the top of the foundation for rebar dowels and secure the dowels in the holes with epoxy. An engineer should be consulted for the size and frequency of the dowels and the minimum thickness of the correcting layer of concrete.

■ Have the framers correct the error with solid-wood shimming at the beginning of the framing process.

Repair mix is rubbed into a section of honeycombing with a spongelike rubber trowel.

CHAPTER 3

Poured Concrete Foundations

THERE ARE MANY types of foundations. In many areas, cinder-block construction is most commonly used, whereas and in other, smaller pockets, wood foundations are the norm. But probably the most widely used foundation type overall is poured concrete. That's because poured concrete is strong, versatile, durable, and moisture-resistant. Also, a poured concrete foundation can be accomplished more quickly with less labor than other systems.

There are many different forming systems for creating a poured concrete wall foundation. The forms may be made of wood, steel, aluminum, fiberglass, composite, or just about any combination thereof. However, the basic principles of erecting, bracing, and pouring remain the same. This chapter explains the process from start to finish using one type of forming system you can buy or rent. I'll also describe how to build your own forms.

Most, although not all, foundations rest on a footing. Since footings are the first part of the job, I'll start by telling you how to make one.

Concrete workers shovel concrete around a corner and down the wall in a hard-to-reach area.

Building Footings

Walking in deep snow is very awkward and strenuous as each foot sinks well below the surface. But throw on a pair of snowshoes and it becomes more recreation than chore as you walk on top. This is the idea behind footings. They distribute the weight of the house over a surface area that is larger than the thickness of the foundation wall, reducing the potential for settling that causes cracks and other problems.

Footings are often included in a foundation design to ensure that most soil types will be able to support the foundation. Let's say, for example, the ground in your foundation hole is loose sand that has a bearing capacity of 2,000 psf. A typical two-story house exerts 3,000 psf on the foundation walls. This means the sand won't support this house on a 12-in.-thick wall. Add a 2-ft.-wide footing under the wall and the weight of the house is spread over twice the surface area. Now the same house is bearing only 1,500 psf, a weight that the soil can easily accommodate (see the illustration top right).

Establishing an accurate wall layout

Before you can build footing forms, you need to establish an accurate layout of the foundation walls. The footing form locations, which don't need to be as precisely accurate as the walls, will be established from the wall layout.

The surveyor will have placed stakes that are offset 10 ft. to 15 ft. from the edge of the foundation so the stakes won't be disturbed by the overdig. The stakes will be marked with the exact amount of the offset. Here's the procedure for laying out the foundation perimeter with string and stakes using the surveyor's stakes as reference.

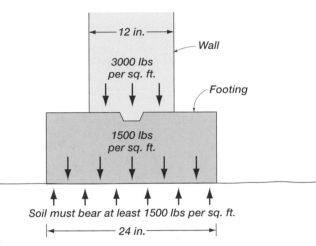

12 in.

Wall

3000 lbs per sq. ft.

Footing

1500 lbs per sq. ft.

Soil must bear at least 1500 lbs per sq. ft.

24 in.

Spreading the load over twice the surface area cuts the bearing weight in half.

Determining the Specifications

You'll need to consider input from several sources when you are determining the height, thickness, reinforcement, and other basic characteristics of your foundation. These sources include the blueprints, building codes, building officials, and sometimes a structural engineer.

A set of blueprints created specifically for the job by a local architect is the most reliable source of information. A local architect will know the local codes and design the foundation accordingly. Specifications for height, thickness, reinforcement, and strength should all be included. If anything was omitted, a quick call to the architect will fill in the blank.

If your plans are mail order or otherwise generic, there's a good chance that the foundation design doesn't conform to local code. You can avoid uncertainty by having a local architect review generic plans. Sometimes the building official will require the blueprints to be redrawn to conform to local code.

Building codes are the minimum standards required by the officials. In some cases, a building official will upgrade a minimum standard to reduce reoccurring problems in a certain area. When reviewing your plan, he may ask you to include footings in your design or perhaps add steel. Site conditions such as poor or wet soil need to be addressed by an engineer who may have to redesign your foundation to account for such problems.

A line is strung from one surveyor's stake to the other. Then the offset for the corner of a foundation is measured from the surveyor's stake, along the line, and located with a plumb bob.

Footing forms are constructed to follow the yellow string, which indicates the outside perimeter of the foundation.

1. ESTABLISH CORNER STAKES FOR ONE SIDE.
Start by stringing a line between two of the offset stakes along one side of the foundation. From each stake, measure in along the string the distance of the offset. At those points on the string, drop a plumb line to the bottom of the hole and mark it with a stake (see the photo above).

2. MEASURE THE LENGTH OF THE SIDE.
Now measure the distance between the two stakes in the hole. That distance should equal the measurement of the foundation on the blueprints. If it is off by only an inch or two, adjust the stakes until the distance is exact. If the distance is off by several inches or more, confer with the surveyor who put in the offset stakes.

3. ESTABLISH THE REMAINING SIDES.
Using the two stakes you put in as reference points, mark out the rest of the foundation and pound in a stake at the location of each corner (see "Laying Out

Laying Out Square Corners

With the corner stakes for one wall accurately placed, you can use the Pythagorean theorem, two tape measures, and a construction calculator to quickly and accurately lay out the rest of the foundation perimeter.

The theorem is written as $A^2 \times B^2 = C^2$ where A and B are the lengths of legs of a right triangle and C is the hypotenuse. Or, in construction terms, A = rise, B = run, and C = the diagonal.

Use a construction calculator to determine all the diagonal measurements and note these on the foundation blueprints (see the illustration below). Then once the first two reference corners are in the ground, have a tape end held at both references and pull the lengths of the predetermined diagonal measurements for a given corner. Where the tapes cross at those measurements is the exact spot of that corner. For example, Corner 3 in the drawing is 31 ft. 2⅞ in. from Reference Corner 1 and 40 ft. from Reference Corner 2. Note that for Corner 4 and Corner 5 in the drawing, one of the tapes is pulled straight along the length of the wall while the other is pulled diagonally.

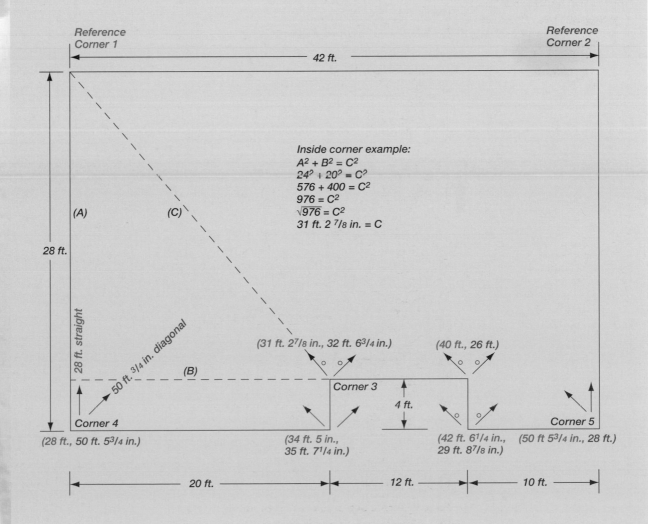

Inside corner example:
$A^2 + B^2 = C^2$
$24^2 + 20^2 = C^2$
$576 + 400 = C^2$
$976 = C^2$
$\sqrt{976} = C^2$
31 ft. 2 ⅞ in. = C

A homemade footing bracket automatically spaces the planks to the correct distance and eliminates a lot of labor pounding in stakes and nailing supports to the planks.

All of the corners of the footing are set up first.

Square Corners," p. 47). Once all of the corner stakes are in, string a continuous line from one stake to the next. What you end up with is an exact outline of the outside perimeter of the foundation (see the photo at the bottom of p. 46).

Sizing the form planks

Footings are typically twice as wide as the foundation wall with a depth that equals the wall thickness. A 10-in.-thick foundation wall will have a footing that is 20 in. wide by 10 in. deep. The widths can increase depending on site conditions and engineering, and the thickness can decrease if steel reinforcement is used.

Once the depth of the footings has been determined, form the footing with 2-by stock that is wider than the required depth. This allows you to shoot a level grade for the top of the footing rather than depending on the foundation hole to be perfectly level, which it never is, and manipulating shorter stock up and down until it is level. I find that 2×12 planks can be used on most jobs.

Forming the footings

You want the foundation wall to be centered across the footing, leaving equal amounts of footing exposed on both sides of the wall. To do this, subtract the thickness of the wall from the width of the footing and divide the answer in half. A 10-in. wall on a 20-in. footing has 5 in. exposed on both sides of the wall.

Use footing brackets to assemble your forms. These brackets are available from concrete-supply companies or they can be easily fabricated from flat steel stock. They eliminate the need for extensive staking or banding. When you drop the brackets over the planks, they are automatically locked together at the correct distance. The brackets should be spaced about every 4 ft. and at every corner (see the photo at the top of facing page).

1. ASSEMBLE THE CORNER FORMS.
Use 16d duplex nails to fasten two planks together to form an outside corner of the footing. In the example, you would position them 5 in. from the string. Next, tack two planks together to make an inside corner of the footing. In the example, set them 20 in. from the outside form planks.

2. POSITION THE CORNER FORMS.
If you are using manufactured footing

brackets, you don't have to tack the inside corner together; just move the inside plank until the bracket fits. The bracket automatically sets the planks to the correct distance. Try to use inside planks that are a foot or two shorter than the length of the outside planks they are facing. This will place the ends of the planks roughly across from each other so that the inside and opposing outside pieces needed to fill in between the corners will be roughly the same size. Repeat this process at each corner of the foundation (see the photo at the bottom of facing page). Use long planks on the long runs and short planks for the shorter walls.

3. FILL IN THE SIDES. Select planks that will bridge the gaps between corners. For gaps of 2 ft. or less, you can use scraps of sheathing. There is no need to cut the planks for an exact fit; just let them over-

The remaining spaces between the corner forms are filled in.

Instead of creating a U-shaped footing beneath the bulkhead walls, a small block is formed.

lap but keep the overlaps short if you can (brackets won't fit over the overlaps).

4. FLY BY SMALL JOGS. For small U-shaped jogs for an item such as a bulkhead, don't bother to follow the layout with the inside plank. Simply form straight across the inside and pour one block (see the photo above). The extra concrete is negligible, but if you figured the concrete closely and you come up short during the pour, you can always steal some from the center of that block area—the rest of the concrete will stay in place.

5. ADJUST THE PLANKS. Once all of the planks are in place, check that the outside planks are all 5 in. away from the string. Then, if you're not using footing brackets, adjust the inside planks to maintain the 20-in. gap between planks.

6. SECURE THE FORMS. The planks do not have to be fastened to the ground, but they do have to be secured in some way to keep them from spreading when

the concrete is poured in. There are a few ways to do this other than using footing brackets.

One way is to pound in steel stakes or rods along the outside of all the inside and outside planks. Locate the stakes opposite each other across the form. These will keep the bottom of the planks from kicking out. If the soil is loose, it may help to backfill dirt against the bottom of the planks. Then drop either prepunched steel or predrilled wood strips over the tops of opposing stakes to keep the top from spreading. Or simply nail wood strips across the top of the planks. If you use nailed wood strips, you can use wooden stakes in place of steel rods (see the photos at top and middle of facing page).

Another method is to thread perforated banding straps beneath the planks and nail the strapping up along the outside face of both planks. At this point you can continue the strapping so that it crosses over the top of the planks. But I prefer to cut the straps off at the sides of the planks and then nail wood strips across the top (see the photo at bottom of facing page). This way, I can cut the wood strips to the exact length (23 in. in this case) so I can use them to gauge the form width in addition to securing the forms in place.

7. REMOVE THE STAKES AND STRING AND BACKFILL. After all of the planks are in place, remove the stakes and string from the inside of the forms and backfill any low spots to keep the concrete from getting under and lifting the planks.

Setting the grade

Foundation holes are almost never level. Using a transit, shoot the top of the footing planks until you find the highest point. Along the inside of the form, measure up the required depth of the

It's well worth spending a little extra time to make your footings as level as possible. A level footing makes it much easier to set the wall forms.

Steel stakes keep the bottom of the forms from separating, while pieces of wood with holes drilled at the correct distance hold the top of the forms in place.

Perforated steel strips can be used instead of wood to hold the top of the forms together during the pour.

Banding straps hold the bottom of the footing planks together, while precut wood strips gauge and secure the top.

A chalkline is snapped between two grade nails located earlier with a transit. Extra grade nails will be added along the line.

footing and tack a nail into the plank. Now shoot that nail grade at every corner and halfway between long runs, tacking a 6d nail at every shot.

If there is a low area in which the grade is above the planks, then you have to raise the planks by shoving dirt under them until the grade can be shot. But if there are more than just one or two low areas, drop the original grade down and reshoot until it fits inside all of the forms. This means, of course, that you have to dig out the inside of the planks at the higher spots to maintain the correct footing depth.

Snap a line from one grade nail to the next, then tack grade nails in along the line every few feet (see the photo above).

Adding steel

If steel reinforcement is required, now is the time to put it in. It's most common to run two lengths of #4 (½-in.-dia.) or #5 (⅝-in.-dia.) rebar in residential footings. As you place the rebar in the forms, it

should be secured about 3 in. away from the inside of the forms and at least 3 in. off the ground. An easy way to hold the rebar in place is by using steel chairs.

Using small bolt cutters, cut the chairs into lengths that are several inches shorter than the width of the forms, in this case about 14 in., and place them in the bottom of the forms about every 4 ft. Lay the rebar near the outside ends, on top of the chairs, and secure it to the chair with a twist of wire. Common wire and pliers work fine, but if you are going to be working on more than just one project, invest a few bucks in a wire-tying tool and preformed wire ties.

Another type of chair requires no cutting. It is like a four-legged spider that the steel sits on top of. To keep the steel from falling off during the pour, it is tied to the chair.

Overlap intersecting ends of rebar at least 30 times its diameter—the #4 rebar should overlap at least 15 in. Wire-tie the overlaps together to keep the continuity of the rebar. Try to avoid joints in a corner.

Some contractors prefer not to spend the extra money for the chairs and instead hang the rebar from wire fastened to the furring strips across the top. If you use this method, cut pieces of rebar to the same length as you would the chairs (14 in.) and tie them across the rebar at 4-ft. intervals. These cross members will keep the long lengths of rebar from spreading out toward the sides of the footing as the concrete is poured in.

Pouring the concrete

Pouring the footings is usually pretty painless because the vertical distance between the footing and the top of the concrete truck allows you to add additional long chutes if needed. Having more chutes means you can pour the concrete into place rather than dragging

Rebar in a footing is wire-tied to a chair to keep it in its proper location during the pour.

A short piece of rebar is tied between two lengths to help keep them from moving during the pour. Instead of using chairs, the lengths of rebar are hung by wires from the wood strips.

The concrete is brought to grade with a shovel, then the surface is troweled smooth.

it along the bottom with shovels. The less shoveling the better, for the workers and for the concrete.

Just before you start pouring, spray a light coat of release oil on the inside surface of the planks so they will come apart easily later. Start the pour with the most difficult areas to reach and work your way out. This will give you more time to work the difficult areas before the concrete sets. This is especially important if you are sent a hot load.

A three-person crew is sufficient for most footing pours. One worker directs the trucks and fills the forms roughly to grade. Another follows behind, vigorously working the top of the concrete with the flat of the shovel. This flattens out the concrete across the forms and brings the "cream" to the top. He adds or removes a shovelful of concrete as needed to get the surface even with all of the grade nails. A third worker follows the second with a trowel, smoothing out the surface of the footing (see the photo at left).

Troweling the surface isn't necessary, but it creates a nice smooth surface that is easy to mark and sweep the dirt from before the wall forms are set up.

Connecting the walls to the footing

A keyway is a small channel along the length of the footings. When the walls are poured on top of the footing, the concrete fills the keyway, creating a lock between wall and footing. Create the keyway after the footing has been troweled by dragging a piece of a 2×3 down the middle of the footing, holding the leading edge up. There should be a rut about 1½ in. deep left in its wake (see the photo at top of p. 56).

In cold-winter climates, keyways may be a problem because they can get filled with ice. If the ice is not chipped out, there will be no connection between the footing and the wall. And if the ice forms after the wall forms are set up, the keyways cannot be reached, so chances are the ice will be left to get poured on.

As an alternative to a keyway, you can insert vertical steel into the footing. Cut #4 or #5 rebar into 3-ft. lengths and make a 90-degree bend about 10 in. to 12 in. from one end, creating a small foot. After the footing is troweled, place the foot of the rebar into the concrete and push it in, using a gentle up-and-down shaking motion to prevent air from being trapped, until it is at least halfway into the footing. Start at a corner and place a bar at least every 4 ft. Place an orange plastic safety cap on top of each rod to prevent severe puncture wounds in the event of a fall near the rods.

Ideally, the vertical rebar should be placed and tied to the horizontal rebar

Working with Release Agents

In the old days, used motor oil was sprayed onto the forms as a release agent to keep concrete from bonding to the panels. Today that practice is frowned upon because it raises environmental, health, and hazardous-waste transportation issues. There are many different types of release agents, but here's what to look for in a release agent for most residential work:

- A water-based, paraffin-based, or vegetable oil-based agent, not diesel fuel
- Nonregulated D.O.T. (Department of Transportation) classification, meaning it's safe to transport
- Noncarcinogenic
- Shouldn't stain the concrete work
- Does not interfere with finish-coat bonding, paint, stain, etc.

Supply yards that carry foundation and flatwork hardware and products usually stock a few different types of release agents. If they don't have what you need, they can easily order it.

Dealing with a Hot Load

Pouring out a truck of concrete takes about 20 to 30 minutes under good circumstances. Imagine pouring out a truck starting at one spot and working your way down the wall. When the truck is empty 30 minutes later, you discover you can stand on the concrete at the starting point. Now try to straighten those walls, float the top, and install the anchor bolts.

You just poured a hot load. The concrete sets up at such an accelerated rate that a job can be ruined. The main cause of a hot load is that there was concrete left in the truck from the previous job and it was mixed in with the new batch for your job. That old concrete is usually at least two to three hours old and has begun to cure. This accelerates the setting time of the new concrete mixed in with it. The more old concrete that was mixed in for reuse, the quicker the whole batch will set.

Outside temperature, slump, water temperature, and accelerator admixtures are all factors in the setting time,

but when combined with old concrete (as measured in hours), their effect is multiplied.

Ideally, you don't want to work with hot loads, but it's something many concrete companies do, so it's best to learn to work with them in case you get stuck with one. The first thing to ask the driver when he arrives is, "Do you have any old?" If he does have a hot load, it's best to wait for another truck with fresh mix. Then pour the hot load, keeping the level well below the top of the finish grade lines. Pour the second truck immediately on top of the old and work the joint with a 2x4 as they meet.

If the second truck is also hot, you may want to send the trucks back. After all, you are paying for new concrete.

If the last truck, the one you use to top off the forms, is a hot load, place, finish, and install the anchor bolts as you go. It will take longer to empty the truck, but the job will be done right.

The leading edge of a piece of 2×4 stock is kept above the concrete as it is dragged along the center of the footing to create a 1½-in. by 3-in. keyway.

The best time to shoot the bottom of the hole for level and lay out the foundation is just after the excavator finishes but before he pulls his machines out. If the hole has to be corrected, it can be done immediately without interfering with your schedule.

Vertical rebar is tied in place as required by the project specifications.

before pouring the footing, and in some areas this may be required. Check with your building official.

Installing Foundation Wall Forms

Most forms will shift and stretch slightly as the concrete is poured. If you set the forms as close to plumb and square as possible, the walls will be well within standard building tolerances, even if the forms do move slightly.

Repeating the layout

Remove the footing form planks and sweep the surface clean of any dirt and concrete debris. To eliminate any accumulated error in the footing layout, it's a good idea to start your wall layout at the same two corners where you started to lay out the footing. For our example 10-in.-thick wall, measure in 5 in. and

mark both corners. Measure the distance between the two marks and adjust them until the distance exactly matches the foundation wall length specified in the blueprints. Now from those two reference points, repeat the layout process you used for the footing and mark all of the foundation corners on the footing. Snap chalklines from one corner to the next until you again have a complete outline of the outside face of the foundation.

Marking the footing

A quick visual reference for the location of the different elements of the foundation is helpful when you load the hole with forms and also helps to prevent oversights. Following the plans, measure along the outline of the foundation and mark elements such as basement windows, beam pockets, and sewer chases. Then highlight the marks with brightly colored spray paint. You should make the marks on the inside 5 in. of the footings

Piece panel sizes and locations of windows and other details such as sewer and water chases are spray-painted onto the footing. Now the crew can easily see how to load and set the foundation walls without referring to the plans.

Building Your Own Forms

When you have plenty of foundation work for your crew to do, it doesn't make economic sense to build your own forms. It's more profitable to buy prefabricated forms and keep the crew on job sites. But if you want to keep the crew busy during down times, setting them to form building can make sense, especially if you can set up an efficient operation like the one shown here.

You'll need to have two types of hardware pieces made up by a local sheet-metal shop. One is a simple sheet-metal plate that reinforces the joints between stiles and rails. It is predrilled for one nail into a rail and one into a stile. The other is a piece that wraps around two sides of the stiles at points where T connectors will join panels to each other. These pieces prevent the Ts from digging through the wood during the tremendous pressure of a pour. They are predrilled for the T connector and for four nails.

Framing members are first precut and stacked, then they are roughly placed in position on an assembly platform.

Using pneumatic framing nailers and spikes, the stiles, headers, and rails are fastened together. Note that the reinforcing plates for the Ts are already in place. These plates are recessed into the top of the stiles to protect them when forms are slid together during stacking.

Reinforcing metal plates are nailed into shallow dadoes in the stiles and to the rails, again for protection during stacking. The rails are reduced in width at each end to meet the stiles flush to the bottom of the dadoes.

The frame is flipped over and ¾-in. high-density overlay plywood is aligned and nailed to it.

A crew member starts to load the hole with panels, while the others lay out the walls on top of the footing.

Loading the holes

Placing all of the forms and hardware down into the hole is called "loading" it. The best way to load most holes is to line the forms along the perimeter of the hole and place the hardware in the center. As the panels are unloaded off the truck, drop each one into the hole against the bank. You should set them in pairs, with the first being dropped facing away from the bank, then the second sliding face in against the first. This way you won't waste time spinning them around into correct position. The piece panels are dropped along the bank in the area of their designation on the footing.

Sometimes a foundation may have a small area that requires many piece or special panels, such as a 45-degree bumpout. Rather than stacking all of these panels against the bank in a bunch to be sorted through again later, carry them into the hole and stack them on the ground in order, near the area where they will be used.

Setting the forms

When setting the forms, begin at the corner farthest from the access ramp/entry point. This keeps your path in and out of the hole clear until the last few panels go up. Connect two deuces at 90 degrees to each other using special corner hardware, then line up the inside faces of the panels along the chalkline on the footing. Place a prebuilt inside corner panel about 10 in. away from, and facing, the outside corner panels. Insert the flat metal T-shaped hardware, called Ts, through the slots along the rails of the panels, then slide foundation rods, flat ties in this case, onto the Ts to bridge the gap between opposite panels.

The ties maintain the two opposing panels at the correct distance apart and keep them at that distance as the con-

so you'll still be able to see them after the forms are up.

The forms shown here are built 2 ft. wide (referred to as deuces), as are many form systems that are premade as opposed to built on site for single use. Panels built smaller than 2 ft. range in 2-in. increments from 2-in. fillers to 22-in. panels. These are referred to as piece panels. So if a foundation wall's length is an even number of full feet— 20 ft. for example—only deuces are needed to create that exact length. But a wall that is 21 ft. 2 in. long will need a pair of 14-in. piece panels in addition to the deuces. Determine the piece panels needed for each wall and spray-paint the panel size in its approximate location on the footing. Now you have a "foundation-at-a-glance" layout on the footing.

crete is poured into the forms. In this example, 10-in. ties are used. They actually measure longer than 10 in. to catch the hardware, but they hold the panels exactly 10 in. apart. If you want 12-in. walls, spread the panels farther apart and use 12-in. ties.

Set the next pair of panels, inside and outside, up against the first corner pair. The Ts are long enough that they extend through the slots in the rails of the second pair. Next, drop in the wedge-shaped metal hardware, called wedges, into the slots on the end of the Ts (at this point just let them stay loose in the slots). This locks the adjacent panels together (see the photo below).

Before you continue setting, plumb the outside corner of the outside forms. Shim beneath the forms if necessary. Having a second form attached to both sides of the outside corner panels helps keep it steady. Plumbing all of the corners this way makes squaring the job faster. In fact, if all of the corners were exactly plumb, the job would already be square after you finish setting (see the photo at the bottom of p. 62).

A flat foundation tie is slipped onto the metal Ts that have been inserted from the opposite side of the rails. The ties space and hold the forms the correct distance apart.

A metal T is inserted through two panels and also through the slot of the flat tie between them. A wedge slides down into a slot at the end of the T to lock the assembly together.

The size of your crew determines your plan of attack. With only two crew members, it's best to lay out the foundation on the footings, completely load the hole, and then begin to set up the panels. If there is a third person, have him start to load the hole as you are marking out the foundation, then give him a hand until about one-third of the hole is loaded. At that point, begin to set the forms as the third person continues loading. If there are more than three in the crew, two should be marking out the foundation while the others load the hole. Once the footings are marked, the two who were marking can get busy setting the forms. Once the hole is completely loaded, the rest of the crew jumps in and starts setting. Once the setting begins, it should continue without interruption.

■ **WORK SAFE**
■ **WORK** SMART
■ **THINKING AHEAD**

Be sure to close the panels up tightly together as you're setting them. Just a ⅛-in. gap between panels on a 16-ft. wall will make your wall 1 in. too long.

Two crew members each take a wall and work away from the first corner; eventually they will meet and "close up" the forms.

The first corner is checked for plumb before proceeding with the rest of the wall. In theory, if all of the corners are perfectly plumb, the job is automatically square and true.

Have two crew members each take a leg of the corner and work away from each other. Repeat the process of inserting Ts, placing flat ties, adding a pair of panels, and lightly locking them with wedges. As panels are added, set the inside face of the outside panel even with the chalkline first, then position the inside panel so that the ties slip on the Ts properly.

Setting temporary bracing

You will notice after setting about 6 ft. of wall that the open end of the formed wall can easily sway in or out. The more you set, the worse it gets until ultimately, if left unchecked, the wall can pull itself over, racking and twisting out of position. To prevent this from happening, set temporary support braces about every 8 ft. as you set the forms.

To keep the forms upright, 2×4 braces are temporarily located opposite each other about every 8 ft. or so.

Eye the open end of the wall so that it's close to plumb. Jam one end of an 8-ft. 2×4 into the ground a few feet away from the footing, and slide the other end under the header of the end panel (see the photo above). Duplicate the brace against the opposite panel, making sure there is pressure against both panels.

Reinforcing with Steel

If rebar will be included in the lower section of the walls, you must place it as you set the forms. Any steel planned for the upper section of walls can be added once all of the forms are set. With some form systems, the outside panels and ties are completely set before the inside panels are put up. If this is the type you are using, once the outside perimeter of

Foundation Rods

Foundation rods are also known as foundation ties, form ties, form rods, snap ties, and snap rods. The type of rod you use depends on the type of form system and the type of job you are setting. Most ties will be strong enough for typical residential work that isn't unusually high or wide. If your foundation plan exceeds the average residential limits of 8 ft. tall by 12 in. thick, check the ratings of the ties you are using in conjunction with the form system. If your forms are homemade, the tie manufacturer will be able to recommend the appropriate tie based on the job specifications and tie spacing.

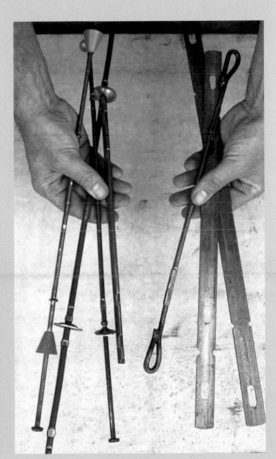

Here is a sampling of various foundation ties for different form systems.

Rebar is bent to a 90-degree corner on a combination rebar bender/cutter.

forms are set, place and secure all rows of the steel before starting the inside panels.

Cutting and bending rebar

A chopsaw or circular saw with a metal-cutting blade works fine to cut rebar. A reciprocating saw with the correct metal-cutting blade will also work, although a little more slowly. Cut the rebar a little more than halfway through, then snap it off to speed up the process. An oxyacetylene torch is another way to cut quickly through rebar.

A manual rebar cutter/bender is the most versatile tool for getting both jobs done. It requires no electricity or fuel and is virtually indestructible. A good cutter/bender will cost from $250 to $350.

Hand-held battery-operated hydraulic rebar cutters are available for around $2,000. Electric rebar benders are also available. These types of units are not cost-effective for residential use; they

are found on large commercial projects and in factory environments.

Placing rebar

If there is a detailed steel plan for the project, follow it exactly. The engineered strength of the wall or structure will only be achieved by constructing it as it was designed.

Securing rebar

A considerable amount of force is introduced during a pour by the mass and movement of the concrete and by mechanical vibrators, if used. Tying the rebar pieces with wire at all intersections and splices and tying it to foundation ties or rods keep the rebar in position throughout the pour.

Looped-end wire ties are sold in different lengths depending on the size and number of rebar to be tied together. Six-inch ties are long enough for residential work that uses #4 or #5 rebar. A wooden-handled twister costs about $5 and easily twists the wire ends tightly against the rebar.

Wrap the looped-end wire tie around the intersection or splice of two pieces of rebar, closing the looped ends together. Insert the hook of the twister into the ends and twist the tool like a New Year's Eve noisemaker. It takes about five seconds. An automatic twister does the same job except a pulling motion on the tool causes its wire hook to spin.

Another method of tying is to use a spool of wire and a set of spring-loaded lineman's pliers. A spool holder attaches to your belt, allowing the spool to rotate freely as the wire is pulled. Loop the wire around the rebar and then twist and cut it using the pliers. There is a hand-held, battery-operated wire-tying tool available. The tool is fed from a spool of wire and automatically wraps the rebar, then ties and cuts the wire in a few seconds. Unfortunately, the tool costs about $2,400, making it cost-effective only for constant production use.

Splicing rebar

Where rebar pieces end, they must be overlapped and secured. The amount of overlap (if not specified on the plans) should be a minimum of 24 times the diameter of the rebar, or at least 12 in. A higher standard of 30 times the diameter is more commonly used. Tie the rebar together close to the end of both members of the overlap. In other words, the overlaps will be tied twice. Any point where rebar intersects, such as the horizontal and vertical members in a wall grid, must also be secured together.

Positioning rebar for corners and angles

For walls or footings that have double horizontal members, one toward the inside of the wall and one toward the outside, there are special considerations

Two pieces of rebar are quickly fastened together with a twister using wire that's premade with loops at both ends.

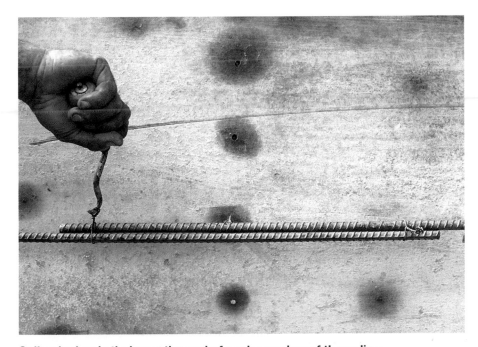

Spliced rebar is tied near the end of each member of the splice.

The length of rebar nearer the inside of the corner continues until it reaches the outer rebar before making the turn.

for how the rebar should be cut and positioned. Adjust the rebar lengths to avoid lapping the outside rebar at the corners or anywhere the foundation is angled. For the inside pieces, do not make the bend near the inside of the corner. Instead, continue the rebar past the inside corner and then bend it where it meets the outside rebar as shown in the photo above. Overlap the inside and outside pieces at least 12 in. Ideally you should tie vertical members at any point where inside pieces cross outside pieces.

Placing rebar below grade

Rebar contained within concrete that is to be in contact with soil should be placed 3 in. away from the soil. Steel chairs are available in many different heights to hold the rebar in place.

Preparing for the Pour

Once you have completed setting all of the panels, you're about half done before you're ready to pour. The rest of the process includes straightening and squaring, bracing, shooting the grade, and installing the incidentals (windows, beam pockets, brick shelf, etc.).

The next step is to install the metal channel that locks the tops of the panels into alignment. But before you start installing channel, slide the basement windows roughly in position between the forms. Once the channel goes on, there won't be enough room.

Installing the channel

Press the 8-ft. lengths of metal channel over the tops of the forms. Start about 8 in. to 10 in. from the corner so the channel ends at least 8 in. over a panel. Leave about a 6-in. gap and start the next piece of channel. At the end of the wall, select a length of channel that will end at least 4 in. away from the end of the corner panel. If there is a piece panel in the wall, you have to select the right size channel so that the butt end doesn't fall over the seam of two panels. It should always fall a good 6 in. to 8 in. away from the seams.

First, install channel along the outside panels, then place channel on the inside panels in the same way except for the layout. Start the inside channel so that the center of an 8-ft. length falls opposite two butt ends of the outside channel.

As you install channel, relocate the 2-by braces so that they are at every channel break along the outside wall. Be sure to place an opposing brace against the inside corresponding panel.

Stringing the walls

The only way to ensure straight walls is to run string from one corner to another and line the forms up directly beneath them.

Tack a 16d duplex nail about halfway in along the outside edge of every corner. If a wall is less than 10 ft. long, don't bother with the string unless you need it for squaring the job; the channel will keep short walls in line. Tie string (nylon mason's twine works best) from one corner nail to another, adjusting the nails in or out until the string falls in line with the inside face of the outside panel of each corner. Pull the string tight so that it is about 2 in. up over all of the panels.

Once all of the walls have been strung, adjust the braces to line the wall up beneath the string. The best way is to have one worker on the top of the wall eyeing the inside face of the panels to the string, and two workers, one inside and one out, resetting the braces as needed.

Squaring the job

First, measure all of the wall lengths. To get an accurate square, the front and rear measurements must equal each other, and the sides must also be the same as each other. The wall lengths should be equal to or a fraction of an inch less than the measurements on the plans. Some wooden form systems are built just slightly smaller than the stated width so that the measurement over a long wall may be ¼ in. to ½ in. small. This is acceptable because these forms usually stretch a little as the concrete is poured

A steel basement window frame is pushed into place between the forms.

Installing lengths of steel channel locks the tops of the forms into alignment with each other.

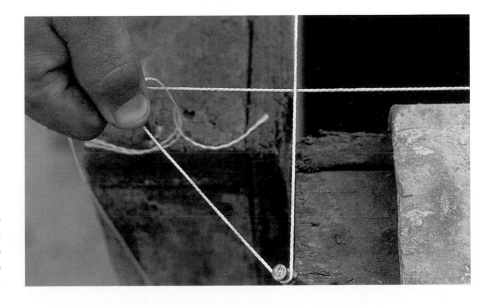

String is set up along the perimeter of the foundation in line with the inside face of the outside panels. This defines the outside of the foundation.

Identical diagonal measurements from corner to corner ensure a square foundation.

When the diagonal measurements are off, the wall is shifted until they read true.

in, and if they don't, it's still preferable to have a foundation undersized than oversized. If a measurement is longer than it should be, check between adjacent panels for small gaps; a few small gaps quickly add up.

Once all of the lengths have been adjusted and are correct, measure diagonally across the largest square or rectangle formed by the strings, then move to the opposite corners and measure the other diagonal of the same area. To be square, the diagonal measurements must be equal. If the diagonals are not equal, subtract one from the other to get the out-of-square distance. To get the diagonals equal, move one wall until the diagonal measurement is adjusted by half the out-of-square distance. You may have to pull the wedges out at certain intervals so that you can move a long wall in smaller sections.

On a complicated foundation layout, start squaring with the largest rectangle. Once it is correct, measure parallel lines off the square for the shorter walls. If you have to adjust any of those walls to get them parallel to the large square, make sure you don't move any walls that will alter the strings of that large square.

Setting grade

The next step is setting the grade to determine the top of the concrete pour. Most jobs specify an 8-ft.-tall wall but actually end up with a wall that's a few inches shorter. The panels are a full 8 ft. tall, but the channel extends at least 2 in. into the form. Other systems that don't use channel can be poured higher but the grade is still kept below the top of the panels.

Sequence for Squaring a Form

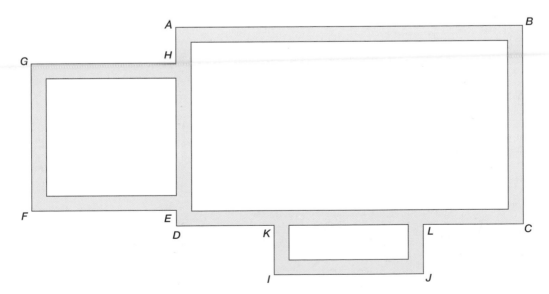

1. Set strings outlining the box ABCD and EFGH and set a string between I and J.
2. Measure all lengths of the walls along the strings and adjust until they are the correct length.
3. Measure diagonals AC and BD. Subtract one from the other to determine the "out of square" length.
4. Move BC until diagonals AC and BD are equal.
5. Measure wall FG to makes sure it is parallel with AD.
6. Measure the diagonals FH and GE. Slide wall FG if necessary to correct.
7. Make sure IJ is parallel to CD. Check diagonals IL and JK. Move wall IJ if necessary to correct for square.

The lead worker uses a laser level to set grade nails at every corner. Crew members follow behind, snapping lines between nails and installing grade nails about every 2 ft. along those lines.

From the top of a panel, measure down 3 in. and tack a 6d nail about halfway in. Then, as you did for the footing grade, on the inside face of the outside panels, shoot the nail with a transit and set grade nails in every corner and in the middle of long wall runs. Snap lines between the nails and tack in grade nails on the line about every 2 ft. to 3 ft. Tack the grade nails in the center of each panel so you will know where to look for it if it gets temporarily buried by concrete during the pour.

Installing Incidentals

Now that the job is set, strung, squared, and the grade is marked, it's time to install the foundation details. It is important to double-check the plans. If there is any confusion or question about a detail, measurement, or possible omission, check with the architect.

Installing basement windows

The most common basement windows are the inexpensive, steel-framed, single-pane units. These are available in different depths to match the thickness of the wall. To install these units, slide the frames into position as measured off the corner of the foundation, and keep the top of the frame about ¼ in. below the grade line. The concrete settles a little after it is poured, and you don't want the top of the frames to rise above the top of the foundation wall. Place a level across the top of the frame to double-check, but if it is out of level and you've followed the grade line, something is seriously wrong with either the level or the grade of your foundation. Secure the frame in place by nailing through the holes provided in the flange into the panels.

Some window companies, such as Andersen, manufacture a series of insulated, wood, awning basement windows. They are about twice the cost of a steel basement window, but the difference in appearance, function, and efficiency is dramatic. Special steel bucks are needed to hold these windows in place while the concrete is being poured, but the bucks are very durable and will last through countless installations. The bucks also form an attractive splay in the concrete around the window. The wooden frame of these windows is temporarily secured to the steel buck, then the unit is set into the wall just like a steel-frame unit. Once the walls are poured, the steel bucks are removed.

Sometimes plans call for a conventional window. In this case, build a pressure-treated wood box of the appropriate size with a 2×2 strip of wood nailed to the outside of these jambs along the sides and bottom. This strip acts as a keyway, locking the box to the concrete. The box is secured in the wall forms by nailing from the outside. During the house framing, the window is installed with the rest in the usual manner.

Creating beam pockets

Small rectangular pockets are created in the walls to support carrying beams for the frame of the house. The dimensions of these beam pockets are not critical as long as they are oversized. Manufactured beam-pocket forms are made of steel and include flanges with predrilled holes for securing to the forms, much like the steel basement windows.

Beam-pocket forms can be constructed from wood by nailing together 2-by stock cut to appropriate size. They can also be fashioned in the same manner out of 2-in. rigid Styrofoam®, which works just as well and is easier to cut and nail together.

To install, hold the wood or Styrofoam beam-pocket forms at the correct location against the face of the inside panel, then use 16d duplex nails through the outside of the form to hold the pocket in place. When you are pouring

After the window frame is leveled, it is nailed to the forms. The top of the window frame is about ¼ in. below the grade chalkline.

A metal beam-pocket form is removed after a pour. The beam-pocket forms are nailed to the inside of the wall forms before the pour.

Blocks of polystyrene are nailed to the inside of the forms that will form a beam pocket in the top of the foundation wall. The polystyrene forms will be popped out after the forms are removed.

When removed from the poured wall, this metal form will leave a hole in the wall for the sewer pipe. It is tapered to allow it to be knocked out of the solid concrete.

concrete, don't pour on top of or right next to the beam pocket. The force of the concrete can knock it out of place, especially a Styrofoam one. Pour a few feet away and move the concrete around the pocket with your shovel.

Whatever type of beam pocket you use, make sure the pockets are 1 in. wider than the beams they are to support and that they provide a bearing shelf that is at least 3 in. deep. Most beam pockets are 4 in. to 5 in. deep.

The tops of beams must be flush with the tops of the sill plates. So, subtract the thickness of the sill plate(s) from the depth of the beam to determine the elevation for the bottom of the beam. Set the bottom of the beam pocket about ½ in. lower than the figured beam elevation. This gives the framer a little wiggle room and allows him to shim up the beam exactly flush with the sill plate(s).

Installing chases

A chase, also called a sleeve, is any hole through the wall in which to allow utilities such as sewer pipes and water lines to pass. The most common chase form is PVC pipe. The diameter of the pipe must be large enough to let the utility pipe easily slide through. For example, if you will be using a 4-in.-dia. sewer pipe for your utility connection, use a 6-in. PVC pipe for the chase. Predrilled flanges are available that slide onto the ends of the chase pipe, making it easy to secure to the forms.

Metal chase forms in the shape of tapered cubes are also available. The taper allows the forms to be knocked out of the wall. Unlike pipe, these forms can't be altered or cut to size on site; they must be purchased for a particular wall thickness. Styrofoam chase forms can be built on site and are easy to place and secure by nailing through the forms. Don't use wooden forms unless you construct a block that is tapered, because they are very difficult to remove from the walls once the concrete has cured.

Regardless of the type of chase form used, be sure to check with the subcontractor that will install the utility through the chase. Some locations are not critical, but others, such as the sewer, may need exact placement.

Making a brick shelf

Anywhere brick veneer will be installed, there must be a shelf created in the foundation to support it. Brick shelves typically extend 4 in. from the foundation wall and are 8 in. to 2 ft. down from the top of the wall. If you arc going to be doing a lot of jobs with brick shelves, it may be worth it to fabricate a wooden brick-shelf system that can be reused. Otherwise, you can quickly, but not necessarily cheaply, make a brick-shelf form of rigid Styrofoam.

The height of the brick shelf should be detailed on the plans. If it isn't, the excavator can give you an idea where the finish grade of the ground will be. The brick shelf should be at that level or a little higher.

Once the height of the brick shelf has been determined, make the forms by ripping 2-in.-thick sheets of 2×8 rigid Styrofoam to a width that matches the height. Make twice as many rips as you'll need to extend along the length of the area to be veneered since the Styrofoam must be doubled to achieve a 4-in.-deep shelf.

Slide the first layer of Styrofoam down against the inside face of the outside panels. When the top of the Styrofoam is even with the grade line, nail it into the forms with 16d duplex nails, sinking the top heads flush with the Styrofoam. This makes the first head of the nail press deep into the Styrofoam to hold it firmly to the forms. Space the nails about 18 in. apart

A shelf is formed on the outside of the wall to support brick veneer.

As the forms are stripped from a recently poured wall, the Styrofoam rips are also removed, creating a brick shelf.

When a brick shelf is called for, shoot the grade on both faces of the forms and tack the nails on the inside panels only. The outside panels now have a reference line for the brick-shelf forms without the nails getting in the way. The inside panels have the grade at which to pour the concrete. The tops of the brick-shelf forms should not be used as a gauge for grade because they can shift and move a little during the pour.

If you are exceeding plan specifications by adding steel to the walls of a foundation, it's not a bad idea to throw a small grid of scrap steel into the piers before they are poured.

Notches are cut on an upside-down piece of polystyrene to allow it to slide into position down over the foundation ties. The first of two pieces is already in place.

and about an inch or two from the top and bottom of the sheet. Once the first layer is fastened, lay the second layer against and even with the first, and secure it in the same fashion by sinking the nail heads into the Styrofoam until you feel them bite into the forms. Make sure to stagger the butt joints.

Sometimes the foundation ties prevent the taller rips from going down into the forms enough. In this case, drop the rips in place on top of the ties. Press down firmly onto the rips until the ties leave a mark. Then flip the rips over and, using an old-fashioned coarse handsaw, cut slots deep enough to allow the rips to slide to the proper elevation. Flip the rips back over and slip them down to the proper elevation. The ties should slide neatly into the slots.

Installing an interior shelf

For certain architectural details, such as a sunken family room, a shelf is needed on the inside of the foundation to support certain floor joists at a lower level than the rest. These are formed in exactly the same way as an exterior brick shelf except that they are placed against the face of the inside panels.

Installing piers

Within the foundation, there must be concrete piers to support the weight under various point loads of the frame usually carried by steel columns in the basement. The size and location of the piers are indicated on the plans and they are generally formed and poured with the footings. But if the foundation is not on a footing, they are dug out by hand and then poured with the walls.

Some building officials will allow you to dig a hole of the right size and then simply fill it with concrete. If you use this method, be sure to dig it a little oversized and keep the bottom of the hole flat. Digging a hole in a cone shape creates a pier that may substantially settle as weight is applied. Other officials prefer to see a hole dug and the pier formed within it.

The top of any pier should be poured lower than the bottom of the basement slab. If it is higher than that grade, when a standard 4-in. basement floor is poured, the floor section over the pier will be less than 4 in. That section will most likely develop cracks because it is supported differently by the pier and it will not cure or move with the same dynamics as the rest of the slab. Pour the piers a few inches lower than the basement grade, and once the support columns have been installed, cover the piers with soil to bring the grade level with the rest of the basement. The soil between the top of the piers and the bottom of the slab helps isolate one from the other, reducing the tendency to crack.

Installing footings for fireplaces

A fireplace footing can be either a large, simple stand-alone pier or an integral part of the wall system. Your plans will detail which type is needed. The pier type is dug out or formed in the same way as any other pier. As with any pier, it doesn't hurt to oversize it and add steel.

If the fireplace footing is formed as part of your wall system, it can be tricky and may be a spot that calls for site-built forms depending on the selection of forms you have to choose from. And, for aesthetics, that wall section that the fireplace sits on is typically poured lower than the rest of the walls so that the brick veneer starts at ground level as shown in the illustration at right. Consult with the mason before the fireplace section is formed. He can often simplify a wall design, saving you a lot of needless fabrication.

Often the piers for the load-bearing columns within the basement are simply holes filled with concrete.

Fireplace Footing

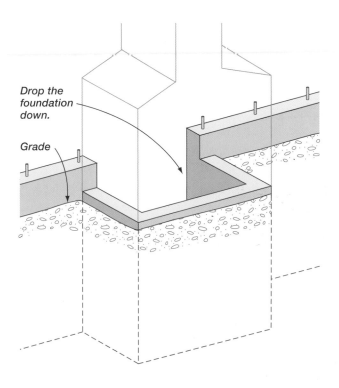

Drop the foundation down.

Grade

Piers for Porches and Decks

Piers that support the point loads of ancillary structures such as decks and porches follow the same basic guidelines and code requirements as poured foundations but are often overlooked and underbuilt. Here's how to plan for and build piers that are not part of the house foundation.

Because of the overdig of the hole, stockpile of dirt, and movement of equipment and trucks, it's almost impossible to pour piers for a porch or attached deck at the same time as the house foundation. However, planning to do a separate pour for the piers can be a problem, because you'll probably only need a yard or two of concrete, which is less than the minimum most ready-mix companies will charge you for. It's also a lot to mix by hand.

The best strategy is to plan to pour the piers during another concrete phase of the job, such as when the basement or garage slabs are poured, or closer to the end of the project if there are walkways or patios to be poured. It will save you time and money.

Check with the local codes for requirements. Most have a minimum diameter, and the size is based on loading and soil conditions. The same frost-line requirements for a foundation also apply to the piers. Often a mechanical fastening system between the pier and the structure is also specified.

Although not always required, a smooth-sided form should be used. If a hole with very irregular sides is filled with concrete, it's not impossible for the frost near the upper part of the pier to grab onto its irregular surface and heave it up. The most common form is the round cardboard construction tube. It comes in different diameters and is sold in long lengths. Once filled with concrete, it is trimmed below grade.

A string to locate deck piers is measured off the foundations and set between two batter boards.

Installing Piers

Lay out the exact positions of the center of the piers using strings and batter boards or offset stakes as described on p. 13. This allows you to remove and reattach the strings a couple of times as needed as you dig holes, install forms, adjust forms, pour concrete, and install hardware.

Excavate the holes deep enough to get below the frost line. If there are a lot of holes to dig, it may be worthwhile to rent a gas-powered auger or a small excavator. Measure the depth of the hole and add the amount that you want to extend above grade. If you want to have all of the tops of the piers to be at the same elevation, make sure to account for any variations in the finish grade.

Cut the tubes to length plus a couple of extra inches. Place them in the holes and carefully backfill, keeping the centers in position.

Using a transit, shoot the desired grade onto each tube and mark it in a few positions around the tube. Now you can either insert nails through the marks and pour until the concrete reaches the nails, or you can cut the tubes at the proper level.

After filling the tubes with concrete, install the fastening hardware. Once the concrete cures, strip away any part of the form that is to be exposed.

Sometimes soil and/or load-bearing conditions require a large surface area bearing on the ground. Instead of using extra-large-diameter tubes or a two-pour footing and tube procedure, you can use Big Foot®, a proprietary system that employs a plastic footing form. The construction tube is fastened to the footing form and then installed and poured as above. This allows the footing and tube to be poured at the same time.

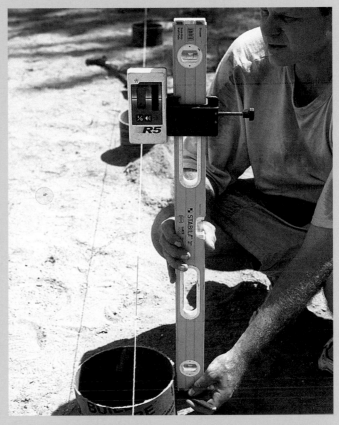

To keep the elevation of the concrete piers the same, the grade is shot with a laser level and marked on each builder's tube form.

Immediately after the concrete is poured inside the tube, the hardware to fasten the deck post to the pier is inserted.

A Big Foot footing system is lowered into a hole. Once in place, a builder's tube form will be slipped over the top and backfilled before the concrete is poured.

To form the void needed for the bulkhead, stops are inserted inside the forms to keep the concrete out of the doorway.

Creating bulkhead openings

Bulkheads provide direct access into the basement from the exterior of the house. Prefabricated bulkhead units include the walls, stairs, and doors. They are made from steel or precast concrete (walls and stairs with steel doors). If you are going to use a prefab unit, all you have to do is provide an opening of the correct width in the foundation wall.

To create an opening, set the wall forms continuously in the normal fashion. Then at the bulkhead location, slide down two stops, one for each side of the opening, the full width of the wall. The stops can be 2-by stock or piece panels of the appropriate width. A 10-in.-thick wall would require a 10-in.-wide piece panel. Panels are better than 2-by stock because the panels are stiffer and offer more nailing area. Plumb the stops with a level, and nail into them from the outside of

Forming a Bulkhead Door Shelf

To ensure that water doesn't find its way under the bottom of the bulkhead door frame and into the basement, a lip should be formed into the top of the bulkhead walls.

Size the walls so that the outside edge is flush to the outside measurements of the bulkhead frame. Cut 2×4s to fit the outside perimeter of the bulkhead walls, then spray release agent on the bottom of the 2×4s. Turn the stock flat and nail into it from the outside of the forms, fastening it in place with the top of the 2x4s about ¾ in. higher than the desired grade. Pour and finish the concrete so it comes halfway up the 2×4s (tack grade nails in the middle of the 2×4s prior to pouring if you don't want to rely on your eye). When the forms are removed, you'll have a ¾-in.-deep shelf to house the door frame.

Depending on the size of the door and the thickness of the wall, you may have to alter the dimensions of the lip. In many cases, the determining factor is the desired width of the bulkhead opening. Functionality and sometimes code require at least a 3-ft.-wide clearance.

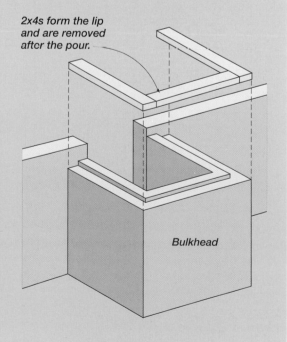

2x4s form the lip and are removed after the pour.

Bulkhead

each panel. Cut 2-by stock and slide it down between the stops to brace them from squeezing toward each other from the force of the concrete. A brace near the bottom, middle, and top of the stops is sufficient.

After the forms are stripped but before the foundation is backfilled, the bulkhead company will set the unit in place, bolt it through the foundation, and seal it against water.

The alternative to the precast unit is to form and pour the bulkhead walls with the rest of the job, and then install the door and stairs later in the project. Because there will be little weight on top, bulkhead walls are usually formed thinner by using shorter foundation ties in that area. The grade is poured lower than the rest of the foundation to avoid a large step up from the ground and so that any rainwater shed from the sides of the bulkhead door can't find its way along the top of the foundation and into the basement.

This is the bulkhead opening created by the stops within the forms.

A prefabricated bulkhead complete with stairs and doors is fastened to the foundation. Notice the through bolts on both sides.

The tops of the bulkhead walls are kept lower than the rest of the foundation. This keeps standing and driven water away from the framing.

Using Alternate Fasteners

Some foundation details make it impossible to use foundation ties or rods either because the detail is too thick or because the joints between opposing forms are not aligned. In these cases, other fasteners are used.

Perforated banding (see p.50) can be used in many odd configurations and can easily be installed at angles and around corners. Where walls are too thick for ties, ¼-in. steel rod called pencil rod can be cut to any length and used instead. Special hardware clamps onto each end of the pencil rod to hold the forms in place. Threaded rod can be used in place of pencil rod. It doesn't require special hardware but is much more expensive.

A foundation detail too large for conventional rods or ties is held in position and reinforced using custom cutting and clamping of ¼-in. steel rod and steel strapping.

Installing drops

In concrete work, a drop refers to a drop in the elevation of the top of the wall. For certain details that are identified in the plans, the grade of the wall drops down to a given elevation. The drops are formed by inserting stops in the forms. The stops extend down to the desired elevation of drop. Two-by stock or piece panels can be used for drop stops. Normally 2-by stock is used because it can be rough-cut to length so it doesn't create an obstacle by protruding far up out of the forms.

After the initial grade is shot, measure down from the top grade line on the inside face of the forms, and mark the distance of the drops as indicated on the plans. Shoot and install grade nails in the areas to be dropped. At the beginning of the drop, slide the drop stops down into the forms until the bottom edge of the stock is even with the lower grade. Use a level to plumb the stock, and secure it with nails through the outside of the forms. Check the nails to make sure they didn't miss or just barely catch the drop. The deeper the drop extends, the more

A stop is checked for level then through-nailed into position from the outside of the forms.

pressure will be exerted on it from the concrete so use plenty of nails.

Fabricating details

Some foundation details may need to be fabricated on site. They will become evident as you plan your panel needs. Two-by stock, ¾-in. plywood, banding strap, ¼-in. steel pencil rod or threaded rod and connectors, and a bucket of ingenuity will take care of just about any detail thrown at you.

A few points to keep in mind are:

- Two-by reinforcement, on edge, should be no farther apart than 2 ft. on center, both vertically and horizontally.
- All ends and corners need vertical 2-by reinforcement.
- Use more fasteners and connectors near the bottom of the forms.

- There is no such thing as too much bracing and reinforcement.

Finishing the setup

Once all of the incidentals are in, install the top run of steel on the uppermost foundation ties. Check the alignment of the forms to the string and rebrace as needed. Now comes the simplest but most important job—checking and tightening.

Check each T to ensure that it is through the slot of the foundation tie. You can check visually or, even better, by grasping the end and wiggling it up and down. Occasionally a tie is knocked off a T just before the next panel is installed. If the tie is resting on top of the T, it appears to be in the right place at first glance. A missing tie is very dangerous. It

The top row of steel is placed and tied once the forms are complete.

At first glance this looks okay, but a closer look reveals that the foundation tie is sitting on top of the T and not interlocked with it. This will allow the forms to bulge and possibly blow apart during the pour.

Oh, No, Blowout!

When you hear a loud pop and look up to see a surprised expression on the concrete truck driver's face, you know you have a blowout. A blowout occurs when a weak spot in the forms, usually caused by missed or failed hardware, breaks under the force of the concrete. The two panels associated with that weak spot burst apart, often distorting the adjacent panels and allowing concrete to come flowing out.

The only way to fix a blowout is to stop the pour, try to block off the flow of concrete a few feet back on either side of the blowout, then dig out the remaining concrete, tear apart, and reset the panels. This can easily add an hour or more to the pour. Meanwhile, the concrete already in the rest of the forms is setting up, which can cause a weakness in the finished walls called a cold joint.

Then there is the labor of moving all of the spilled concrete away from the nearby forms and, if the blowout happened on the inside, removing it from the basement.

will allow the forms to gradually bow out during the pour. If the bow isn't noticed, the panels will surely blow apart at that weak spot, especially if the missing tie is at the bottom of the form. As you check each T, tap its wedge down to tighten the hardware and the panels together.

Applying the release agent

A 3-gallon, heavy-duty pump sprayer is the least expensive method of applying release agent. A small plastic yard sprayer can't supply the air pressure needed to apply the agent in a fine spray, especially in colder weather when the agent thickens.

Certain concrete standards state that you should apply release agent to the forms without getting any on the rebar. But studies have shown that coating the

A fine coating of release agent is sprayed onto the faces of the forms just before the pour.

rebar with release agent does not cause loss of bonding or other adverse effects. If you want to err on the side of caution, spray or roll the release agent onto each panel individually as you set the forms. Otherwise, spray a light coating of release agent down into the wall assembly, completely covering the inside faces of the forms just before you start pouring the concrete.

Planning a pouring strategy

Any site-fabricated details or drops in the foundation of more than 8 in. require a stiff concrete mix (about a 3 or 4 slump) to reduce the amount and duration of pressure on the area. The trucks usually arrive with a stiff mix that you will want to loosen up a little (so it will flow better) by having the driver add water. Once loose, the mix can't be stiffened again, so you have to plan to pour the areas that need the stiffer mix first.

Pour site-built details in sections called lifts. Fill the forms about one-third full and move on to another area. When the second truck arrives, pour another third on top of the first lift. When the final truck arrives, pour the final lift to grade. The intervals between lifts allow the concrete to begin setting up, drastically reducing the pressure it exerts. Between pouring the lifts, make sure to occasionally poke the concrete with a 2×4 to keep it somewhat plastic on top. This allows the next lift to mix with the previous one, preventing cold joints.

Next, keeping the mix stiff, pour into the forms on the high side of the drops. When the level of the concrete rises a few inches above the bottom of the drop, stop and move on to the next one. Once all of the drops have been poured, you can loosen the concrete to about a 5 slump.

The concrete is kept stiff as the drop is filled to within a few inches above its bottom. Later in the pour, after the drop has started to set, the wall can be filled up to grade.

■ WORK SAFE
■ WORK SMART
■ THINKING AHEAD

Work your way around the foundation, pouring it in 2-ft. to 3-ft. layers. Pouring the entire foundation in lifts significantly reduces the force of the concrete pushing the forms out, which in turn, reduces the chance of a dangerous blowout.

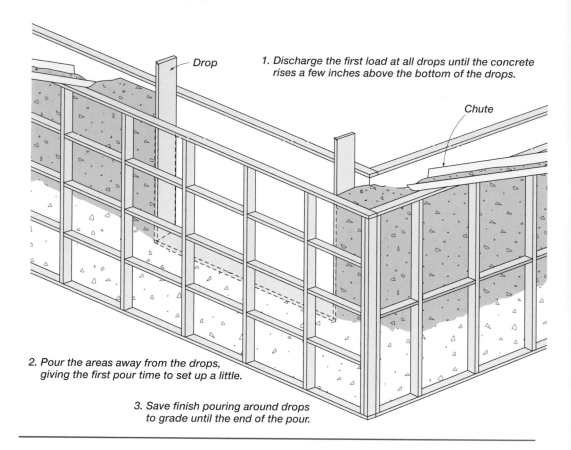

Pouring Sequence for Drops

Drop

Chute

1. Discharge the first load at all drops until the concrete rises a few inches above the bottom of the drops.

2. Pour the areas away from the drops, giving the first pour time to set up a little.

3. Save finish pouring around drops to grade until the end of the pour.

■ **WORK SAFE**
■ **WORK SMART**
■ **THINKING AHEAD**

When pouring areas that you had to fabricate on site, it's smart to have a couple of extra pairs of eyes on the ground watching the forms. Have one person watch the inside of the forms and the other watch the outside. Often a weakness in the forms will become evident as they slowly begin to separate. A sharp eye will catch this, and the pour can be stopped before a blowout occurs.

As explained in chapter 2, concrete sets up fast, leaving you a small window of time to work it, so pour the most difficult-to-reach and time-consuming areas next. Bring the concrete up to grade and work your way out of the difficult areas. Finally, pour the rest of the forms up to grade.

Pouring techniques

A number of things can go wrong during a pour, turning even a simple job into a disaster. Pouring the concrete so that it hits the ties as it falls into the forms will cause the aggregate to separate from the mix, resulting in defects in the wall. As the truck pulls up to the forms, direct the driver until the end of the chute is positioned so that the discharging concrete avoids hitting the foundation ties. You can also use the back of a shovel to direct the concrete away from the ties as it falls.

LIMIT THE FALL. Ideally, to prevent the aggregate from separating, concrete should not fall more than 5 ft. Try to start discharging in one spot, which will quickly build up a few feet, and continue around from that area instead of starting new at another section.

POKE THE CONCRETE. If you pour in lifts or if there is lag time in between truck deliveries, poke the top 12 in. or so with a 2×4 to keep it plastic and workable. When you pour another lift or have a late truck, follow the new concrete around, poking and mixing it into the older concrete to prevent a cold joint.

DON'T POUR ON INCIDENTALS.
Don't pour on top of or right next to any incidental forms such as beam pockets, chases, and brick shelves. Pour near them, and have the truck driver slow the rate of discharge. Use your shovel to guide the concrete into position and carefully watch for any shifting.

WATCH ALIGNMENT. Keep an eye on the wall alignment. As the concrete fills the forms, the walls begin to lean one way or the other. Stop the pour and realign the walls to the strings. Many times there is a short wait between deliveries, a perfect opportunity to check all of the walls.

LOOK FOR BLOWOUTS. Watch and listen for blowouts by constantly scanning the inside and outside forms in the area you are pouring. Look for any bellies in the wall or any deformed ties along the top row (if a lower tie is missed, the concrete forces the bottom of the wall apart and the top of the wall together, causing the top tie to start bending). If the concrete hasn't been poured too high at this point, the panels may not necessarily blow apart yet. A popping noise can be a defective tie breaking under pressure, but that is very rare.

If caught early enough, you can avoid a complete blowout by stopping the pour and bracing and securing the belly from both sides. Then wait until the very end of the job, giving the concrete that is already in there time to set up, before filling that section to grade.

Finishing off

To finish the job, float the top of the concrete with the back of the shovel, adding or removing a touch of concrete until it covers the grade nails halfway. Depending on the size of your crew, try to do this soon after the concrete leaves the chute.

The chute is placed so the concrete does not hit the ties as it is being discharged.

Poking the concrete with a 2×4 during the pour helps prevent voids and other defects, particularly at the corners, stops, and drops.

Insulating concrete forms (ICF) provide insulated basements ready for installation of interior drywall.

Insulating Concrete Forms.

Insulating concrete forms (ICFs) are becoming an increasingly popular alternative to traditional forms for poured concrete. Except for a few brands, ICFs are made of rigid polystyrene insulation, the same material used to make coffee cups. Once filled with concrete, these forms stay in place to provide an insulated foundation. An ICF founda-

tion offers all of the structural benefits of a poured foundation but with much improved insulating properties plus a reduced likelihood of moisture problems.

To the builder, the biggest advantage is that ICFs eliminate the need to transport expensive forms to and from the job site. Once the ICFs are delivered to the site,

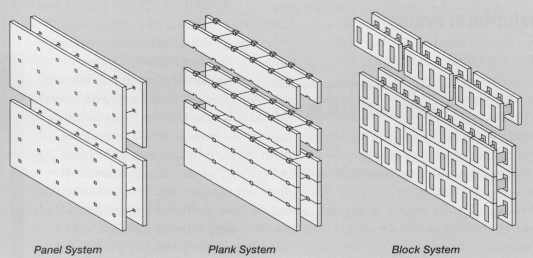

Panel System   *Plank System*   *Block System*

it's just a matter of setting, staging, and pouring. There are no heavy forms to remove or trucks to load. In addition, ICFs have been refined and improved over the past two decades so that now they are easy to learn to use and can be installed with few tools and little equipment.

ICF units come in three types: block, which is most commonly used; panel; and plank. These units are manufactured to form four different wall configurations. These configurations include post and beam, waffle grid, screen grid, and flat wall. In general, block units and panel units are made to form all wall types, whereas I've only seen planks available for flat walls.

The engineering requirements can vary between wall configurations, so make sure you use the correct specifications (steel-reinforcement tables, etc.) for a particular design.

There is some variation in the cost and installation procedures for the different ICF unit designs. But the basic procedures for erecting and pouring the foundations are essentially the same. The four wall configurations are:

■ Flat wall. The flat-wall ICFs form the concrete into a traditional solid flat wall. Both surfaces (inside and out) are flat, maintaining the full thickness of the wall throughout. This type uses

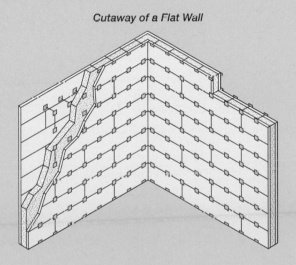

Cutaway of a Flat Wall

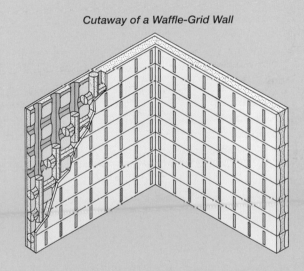

Cutaway of a Waffle-Grid Wall

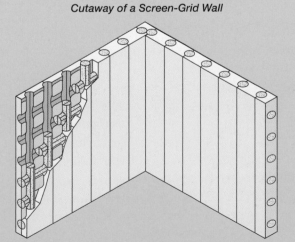

Cutaway of a Screen-Grid Wall

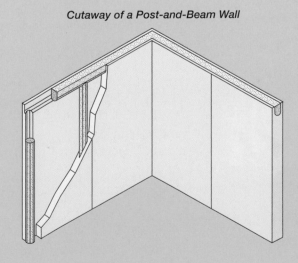

Cutaway of a Post-and-Beam Wall

Vertical steel is placed in a freshly poured footing. The size and placement are obtained from steel-reinforcement tables for ICFs.

(continued from p. 89)

more concrete than the others, but less reinforcing steel is required in the wall design. Many builders who are used to regular poured foundation walls are most comfortable with the flat-wall type of ICF. Here I'll discuss how to install a flat wall using ICF block units.

▪ Waffle-grid wall. The waffle-grid system forms a huge waffle-shaped concrete wall within the ICF. It is also referred to as an uninterrupted grid.

▪ Screen-grid wall. This system forms a similar pattern except that the thin, flat areas of the waffle are occupied with the foam of the form. In other words, the flat section is void of concrete. This is also known as an interrupted grid.

▪ Post-and-beam wall. This type of ICF system creates a post-and-beam structure with the concrete. The size and spacing of the posts and beams are important factors in the bearing requirements.

Installing Footings for ICFs

Lay out and pour the footings as discussed in "Building Footings" on p. 45. Use rebar dowels in place of a keyway, matching the size and spacing to the vertical steel requirements obtained from the reinforcement tables found in "The Prescriptive Method for Insulating Concrete Forms in Residential Construction" published by the Department of Housing and Urban Development (HUD) and available online. Spend extra time making sure your footings are poured dead level. This way you will avoid any extra labor trying to get the first course of ICFs level.

Laying Out the Foundation

Once the footings are stripped and swept clean, snap lines to identify the outside perimeter of the foundation as described in "Repeating the Layout" on p. 57.

Many flat-wall ICF forms are made with 2½-in.-thick foam stock. In this case, the easiest way to hold the first course in place is with 2½-in. light-gauge steel track that is normally used with a steel-stud wall system. Align the outside edge of the steel track to the perimeter lines and fasten it to the footing with 1-in. masonry nails. Four nails per 10-ft. section are sufficient. On some types of forms, the interior corners have a radius, so you may have to leave the track a few inches short of the foundation corners to accommodate this.

Installing Block Units

Once the entire track is in place, you can start the first course by pushing the outside face of the form forcefully into the track.

Set the corner blocks first, then work your way from the corners in toward the center of each wall. Cut the last block to size to complete the row. It's crucial to set the first course level so that the rest of the form installation will go up plumb and level. If the grade of the footings is questionable, shoot the footings with a transit to determine if there is a problem. If the footings are within ¼ in. level overall, you won't have any problems. If they are up to ¾ in. out, you have some shimming and adjusting to do. If they are over ¾ in. out, you have to start ripping the forms to fit level.

Forming Openings for Doors and Windows

To create a rough opening and a sufficient nailing surface for a door or window, you must build a door or window buck and insert

A 2⅝-in. light-gauge steel track typically used for steel-wall construction is nailed to the footing along the lines snapped for the outside perimeter of the foundation.

Installing a Window Buck

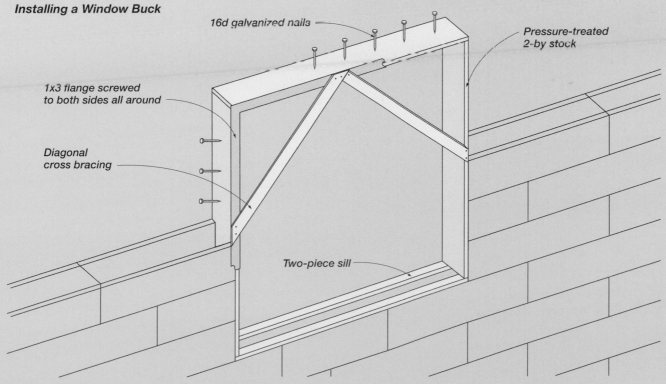

16d galvanized nails

Pressure-treated 2-by stock

1x3 flange screwed to both sides all around

Diagonal cross bracing

Two-piece sill

(continued from p. 91)

it into the wall system. Build bucks from pressure-treated lumber unless the wood will be isolated from the concrete by some type of membrane.

Make the tops and sides of the bucks from 2-by stock ripped to the width of the forms. Make the sills of the bucks from two pieces, leaving a gap between to pour concrete through. To hold the bucks in place during the pour, surround the fronts and backs of the bucks with 1×3 flanges. Square the bucks and nail on temporary cross–bracing to keep them square. To key into the concrete, drive a staggered row of galvanized 16d nails around the perimeter of each buck.

Installing Steel

The horizontal rows of steel are placed as the courses of forms are set. Follow the reinforcement tables for placing steel. Vertical steel is usually put in place after the top course of forms is finished. Cut the steel so that it is a few inches short of the top. In some seismic areas, the vertical steel is cut 6 in. to 12 in. long and then bent at a 90-degree angle at the proper height. Mechanically fasten the vertical steel to the top row of horizontal steel with wire ties, clips, or plastic zip ties.

Stringing and Bracing the Wall

Screw wood or metal scrap stock to the outside of the form on both edges of every corner. The stock should rise above the top course by a couple of inches. Run string from corner to corner, fastening it to the stock so that you have an outline of the outside of the foundation a couple of inches above the top.

Brace the walls according to the manufacturer's instructions. Some require more bracing than others. Common to all types of forms are the diagonal braces that occur every 6 ft. to 8 ft. These allow you to adjust the top of the walls straight. They can be attached to vertical braces against the wall or to horizontal braces fastened along the top course.

String is tied to a scrap piece of wood fastened to the corner of the foundation. The string is aligned to the outside face of each corner, providing a guide for aligning the rest of the wall.

Make sure that the wall section attached to each diagonal brace is directly beneath the string, then fasten it securely.

Installing Staging

ICF forms do not provide a solid, safe platform from which to pour concrete. Workers should not "walk the walls" as they do with wood or metal wall forms. Whether the concrete is poured from a truck or pumped, there must be access for the workers at the top. There are many types of commercial staging and scaffolding systems available for purchase and for rent. Some ICF manufacturers recommend certain types or even have their own system for sale.

Pouring Concrete

To pour concrete into an ICF foundation, the chute of the concrete truck must be able to be placed directly over the top of the forms all the way around. It's very rare to have a site with such complete access, so more often than not, an ICF foundation is pumped.

Start the pour at one spot and pour only about one-third of the forms' height—typically about 3 ft. Continue pouring to this height all the way around the foundation. When you get to the starting point, pour another one-third all the way around. By the time you begin the final one-third of the pour, the first has begun to set (change from a plastic to a solid state). This greatly reduces the amount of fluid pressure at the bottom of the walls where it is greatest.

At the completion of each lift, check the wall alignment to the strings. If the forms are within 1/8 in. of the string, then leave them until the end of the next lift. If they are out of alignment more than 1/8 in., set them back into position.

Once the pour and all work has been completed, check the wall alignment one last time and tweak it straight so it is lined up with the string.

The exterior face is carefully aligned with the string above it. Continuous staging is set up around the entire perimeter within the walls to provide a working platform during the pour.

CHAPTER 4

Crawl Spaces, Retaining Walls, and Other Variations

THERE ARE MANY variations to the basic poured concrete foundation described in the previous chapter. In this chapter, you'll learn about the most common, including frost walls, crawl spaces, and slabs on grade. You'll also learn about foundations for additions, which have their own set of considerations and obstacles.

I'll discuss proper steel placement and pumping concrete—two issues you are bound to run into sooner or later when doing residential concrete projects. Finally, I'll cover retaining walls, since they are, for the most part, just reinforced foundation walls with a couple of extra details.

A foundation for an addition can often be more involved than one for a new house.

Frost Wall

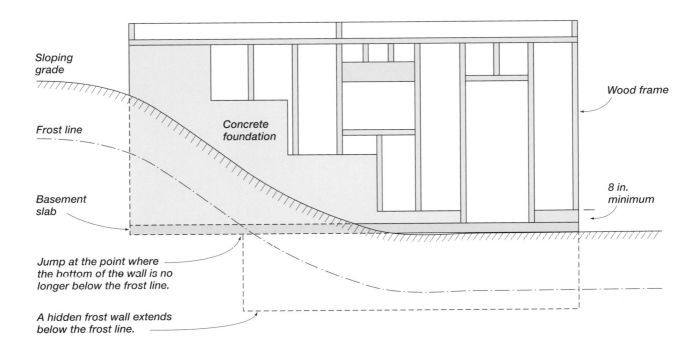

Sloping grade

Frost line

Basement slab

Wood frame

Concrete foundation

8 in. minimum

Jump at the point where the bottom of the wall is no longer below the frost line.

A hidden frost wall extends below the frost line.

Frost Walls

In cold climates, moisture in soil near the surface of the earth freezes. As it freezes into frost, the moisture expands, moving and lifting the soil and anything on top of it. If frost gets beneath a foundation, it can lift it, causing it to crack.

A frost wall is any part of a foundation wall that is created for the sole purpose of extending the bottom of the wall below the frost line. The depth of the frost wall is established in regional tables in the building codes. Ask your building official for the frost-wall depth in your region.

Walk-out basement frost walls

When the land slopes from front to back, and you don't want a lot of the foundation left exposed, you have two options.

The first is to build a full-height foundation. This requires bringing in an enormous amount of dirt to cover most of the exposed foundation. The second option is to step the foundation down to match the land, so that at the back of the building, the foundation wall is flush to the basement slab on the inside and just the required 8 in. above the existing grade on the outside (see the illustration above).

Stepping the foundation down allows the rear wall to be framed with wood, making it easy to incorporate standard windows as well as standard doors that allow you walk directly out of the basement. This is a popular option where the land lends itself to such a design, because the basement can more easily be turned into useful living space with natural sunlight.

The foundation steps down at the back of this house, creating a walk-out basement that allows for full-sized doors.

In a walk-out foundation design, there is a point where the land slopes down enough so that the bottom of the foundation hole is no longer below the frost line. At this point, the bottom of the foundation must drop down below the frost line. To extend the foundation below the frost line, you need to pour a frost wall. To form the frost wall, the earth is excavated and the forms are set lower than the bottom of the foundation hole. The excavation is a simple trench about 6 ft. to 8 ft. wide, enough to contain the foundation and a few feet on both sides to work in.

The change in elevation of the bottom of the wall is referred to as a "jump," so as not to be confused with the elevation changes along the top of the wall known as "drops." In the example shown in the illustration on p. 95, there is only one jump. This jump is deep enough so that the wall will still extend below the frost line when the slope reaches its lowest elevation at the back of the house. If the house and the slope are long enough, it may be practical to have two or more jumps to save concrete. Although there is only one jump in the example, there are three drops to follow the contour of the land. You can incorporate as many drops as you like as long as the top of the foundation always remains at least 8 in. above grade.

Jumps

When forming a wall with jumps, you want to have a joint between panels as close as possible to the jump. Use piece panels as necessary to accomplish this. Since jump banks are never perfectly plumb, the form must overhang the panels by a few inches if you are setting the higher wall first. If you are setting the

Forming a Drop

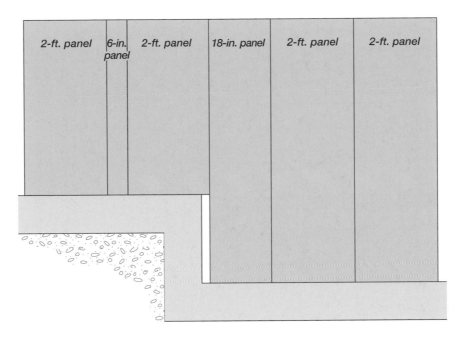

| 2-ft. panel | 6-in. panel | 2-ft. panel | 18-in. panel | 2-ft. panel | 2-ft. panel |

When lining up panels at a jump where the bottom of the wall changes elevation, use hook rods in place of foundation ties because the normal hardware usually doesn't line up at these areas.

Voids between the panels and the bank are patched with sheathing scraps to keep the concrete from seeping out during the pour. Note the clamps that join two panels over the jump.

lower portion first, bring it as close as possible to the bank.

Insert special rods called jump rods or hook rods into the lowest section of the form that ends next to the trench wall. Insert wedges into the jump rods to hold them in place.

Next, if the hardware holes in the upper and lower panels line up to each other, use the standard method of inserting Ts, rods, and wedges to hold the assembly together. If they don't line up, use the jump rods and wedges in all of the hardware holes to keep the forms at the correct width. Then fasten the upper and lower forms together using studs or banding. You can also use hardware called clamps that are specially fabricated for this purpose.

Patch the gaps left between the bottom forms and the wall of the trench with plywood scraps nailed into the side of the forms, then backfill against the plywood to keep it in place during the pour. Keep an eye out for movement around the patched areas as you pour the concrete.

Step-down walls

Step-down or partial walls are the same as walk-outs except they don't step down all the way to the basement slab to create a full walk-out. They follow the same requirements for extending below the frost line. For example, say the existing grade drops 4 ft. from front to back. To take advantage of this, you could drop the height of the foundation wall by 4 ft. around the back side. Then you could frame that missing section of foundation with standard windows.

Crawl Spaces

Crawl spaces are simply very short basements. They can have all of the same details as full-height basements only with shorter walls. The biggest difference is in the windows and access.

Crawl-space windows and vents

In crawl spaces, screened vent units with sliding louvers are typically substituted for windows. These units are made in different sizes to match the wall thickness

Garage Foundations

Since garages have slab floors, not basements, their foundation walls only need to go down to the local frost depth, not the 8-ft. depth typical for basement walls. This is true whether the garage is attached or stand-alone. So, by definition, garage foundation walls are frost walls.

Because of the overdig, however, when attached to a basement foundation, the garage walls begin against the foundation at the same 8-ft. or so height. But once past the overdig, the bottom elevation of the walls jumps up to the required frost-wall depth, saving labor and concrete.

When excavating for a garage, only a trench is needed because unlike a basement or crawl space, the interior is brought back up to grade to accept a slab that will transition to a driveway. The garage slab is poured within the walls of the garage foundation and is gradually pitched down a few inches from the rear of the garage down to the overhead doors so any liquids will flow out of the garage.

Openings in the foundation must be made wherever entry doors or overhead garage doors will be located. This is to allow the slab to meet the outside grade. These openings are created by installing drops inside the forms before the pour. The locations of the drops are determined by the location and widths of the doors. An opening must account for the width of the door plus its framing and trim—normally about 6 in. wider than the door unit. A typical overhead door is 8 ft. wide so the opening in the foundation is 8 ft. 6 in.

The drops should extend at least 6 in. lower than the top of the future slab. If the slab elevation is in question when the drops are inserted, place the drops deep enough so you know they will be well below the bottom of the slab. The extra depth of the drop will be filled in during the pour of the slab (see "Pouring Garage Floors" on p. 162).

and vent-opening requirements, and they are installed the same way as windows. They typically are placed to provide cross ventilation. When standard steel buck basement window units are used, they often are installed backwards so they can be operated from the outside and used for access.

Crawl-space access

Building codes have minimum requirements for the size of access openings into crawl spaces. These requirements can vary depending on whether any mechanical systems are located in the space. Check with your local building official for the minimum required opening in your area.

To form an access opening, start by ripping pressure-treated 2-by stock to match the wall thickness. Nail a frame together that consists of two vertical sides and one horizontal bottom member. The ends of the vertical members should extend over the butt ends of the horizontal member. Be sure to account for the frame and trim thickness in determining the correct dimensions of the frame. The finished opening must be at least as big as required by code.

After you assemble the frames, nail 20d galvanized spikes into the face of the vertical members that will be against the

A vent unit is installed in a crawl-space foundation for an addition.

concrete, stopping just before they protrude through the frame. Stagger the nails and space them about 8 in. apart. This will hold the frame permanently in place once the concrete cures around the nails. Lag bolts can also be used for this purpose.

Slide the frame into place within the forms, setting the tops of the vertical members about ¼ in. below the finish-grade chalkline. Plumb the unit and nail through the forms into the frame with 16d duplex spikes.

When pouring, make sure the concrete fills completely up under the bottom of the frame. Tap the sides of the forms in that area with a rubber mallet to ensure no voids are created.

Slab-on-Grade Foundations

Even though it may sometimes look like it's sitting on the finish grade, a concrete slab that will bear the weight of a structure should always have a foundation under its load-bearing areas. The depth of that foundation is determined by the local frost-wall codes and/or how far down you have to dig to reach bearing soil.

In climates where frost is not a consideration and good bearing soil is within a foot or two of the surface, a slab and its supporting foundation are often formed and poured as one unit called slab-on-

Framing a Crawl Space Access

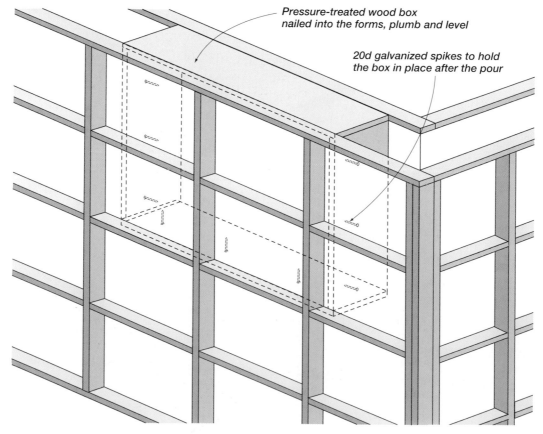

Pressure-treated wood box nailed into the forms, plumb and level

20d galvanized spikes to hold the box in place after the pour

Once the wall is poured and the forms are stripped, an access door is built within the pressure-treated frame.

grade foundations or monolithic slabs (see the illustration at right).

This section explains the extra steps involved in pouring a monolithic slab but does not cover the entire process of forming and pouring concrete slabs in general; that is covered in chapter 6.

Trenching the slab

After the excavator has removed the top-soil and filled back up to the desired height with good compacted soil, he digs a trench about a foot or so wide around the perimeter of the slab. The trench is dug just deep enough to reach good bearing soil and to satisfy any frost-wall codes. (Note: if you have questionable soil conditions as explained in chapter 1, consult an engineer regarding the configuration of the trench and design of the slab.) A trench is also dug in the areas within the slab where there will be bearing partitions or point loads as determined by the plans. Since the inside trenches are inside a heated building, they do not have to meet any frost-wall requirements.

Forming the slab

The perimeter trench in a monolithic slab makes setting the slab form boards

Slab-on-Grade Foundation

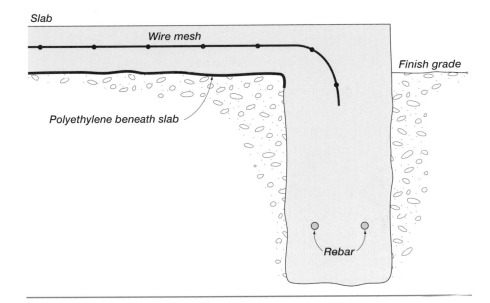

in place on the ground impossible. Instead, they have to be suspended and held in place by horizontal support boards set perpendicular to the outside of the form boards (see the illustration below). These support boards, in turn, are held in place by pairs of stakes.

At each corner and about 4 ft. on center in between, drive a pair of stakes into the ground. The first stake should be

Forming a Slab-on-Grade Foundation

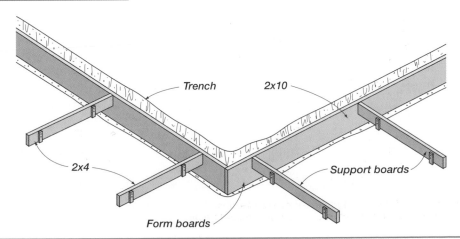

Steel Stakes

Wooden stakes can be a nuisance because they often split and break. If you are going to be working with concrete more than a few times a year, it's worthwhile to invest in steel construction stakes. They are available in various lengths and are predrilled every few inches for fasteners to slip through and attach to the forms.

Steel construction stakes are much more durable than wooden stakes.

as close to the trench as possible without disturbing the trench wall. The second should be a few feet behind the first.

Using the surveyor's layout stakes or your own batter boards, run strings outlining the exact perimeter of the slab. Next, cut 2-by stock roughly into 4-ft. lengths. Establish the finish slab grade with a transit, then shoot and mark each stake. Line up the top of the support boards even with the mark on the rear stake and about ⅛ in. higher than the mark on the front stake. Fasten the board to the stakes, making sure to maintain the 1½ in. away from the string. Now hold the form boards against the supports. Flush up the tops, then fasten the boards to the supports by screwing through the forms into the supports.

To prevent the concrete from undermining and lifting the form boards, the boards should be tall enough to just about come in contact with the ground (2×10s are usually sufficient). Cut the form boards where necessary so they butt over the support boards.

Assemble forms with screws instead of nails so you don't knock them out of place.

Once the forming is complete, use a transit to check the level of the forms where they meet with the support boards. Tweak them into the right elevation by lightly tapping down the front stakes. If there are any low spots (there is always someone who uses the forms as a step), shim up the support board at the front stake with 2-by stock.

Adding the rebar

Install the rebar in the trench according to the plans. If the plans don't call for rebar, it's a good idea to include it anyway. A double row of #4 rebar near the bottom is standard practice in most areas. As with the footings discussed in chapter 3, set the rebar on steel chairs or some other type of supporting system that keeps the rebar in place about 3 in. off the ground as the concrete is being poured (see "Adding Steel" on p. 52). Finish prepping and pouring the slab as detailed in chapter 6.

Once the forms are set, they are tweaked to exact grade by tapping with a hammer and checking with a transit.

Foundations for Additions

Tackling a foundation for an addition often involves just as many details as a new-house foundation plus the extra challenge of dealing with the existing structure and the obstacles surrounding it such as utilities, fences, and flower gardens to name a few. To arrive at an accurate budget of time and expense, you'll need to scrutinize the plans and spend some time at the job site methodically considering how excavation machinery, forms, and concrete will get to the site. If you have to move forms or concrete from a distance, it can cost hundreds of dollars more than if you can bring form and concrete trucks right up to the hole.

Considering the existing foundation

The type of existing foundation is an important consideration in costing the foundation for an addition. Marrying the new foundation to some types of existing foundations is more time consuming than others. Here is an overview of the types of existing foundations you are likely to encounter. Later in this chapter I'll get into the specifics of how to tie the old and new together.

POURED CONCRETE FOUNDATION AND A CONCRETE MASONRY UNIT (CINDER BLOCK) FOUNDATION.

These are the easiest to work with. They are usually plumb and flat, so matching

Planning Ahead

A low tree limb is cut to allow the concrete truck access on site.

A number of considerations must be addressed and accounted for in order to create an accurate budget of time and expense. Visit the job site, scrutinize the plans, and be methodical in determining and assessing the encumbrances.

■ **Access.** Getting the forms and concrete trucks right up to the hole or having to move the forms and the concrete from a distance can mean the difference of hundreds of dollars.

■ **Trees.** There must be a clear path from the road to the foundation hole. If there are trees that the concrete truck can't maneuver around, they will have to go or plan on hiring a concrete pumping contractor.

■ **Limbs.** Many people will look everywhere for obstacles—except up. You can have the forms set, oiled, inspected, and ready to go. Then the concrete truck pulls up and can't get past an overhanging limb. Now you have a crew waiting around and a truck full of concrete. And at this point, you own the concrete.

■ **Overhead utilities.** Communication lines normally hang the lowest and can be propped high enough by a wooden pole for the truck to pass underneath. If the power lines are too low, it's best to avoid the problem entirely and plan for a concrete pump.

■ **Driveway.** There is no way to predetermine whether a driveway will hold up under that 40-ton concrete truck. It is quite possible that the driveway will crack, so it is best to keep the truck off it. If you have no choice, explain the risk to the property owner and have him sign a waiver. Most concrete companies will ask you to sign a waiver before driving a truck onto a driveway.

■ **Sewer pipes.** Some sewer pipes are buried only a foot or two underground. They will crush when a concrete truck rolls over them. If the existing foundation is a full basement or a crawl space, you can see at what level the sewer pipe leaves the house. If it's a few feet below grade, there shouldn't be a problem. If it is any higher, put wide planks on the ground over the pipe in the path of the concrete truck to help disperse the weight.

■ **Underground utilities.** Underground containers and voids such as septic systems, tanks, and dry wells must also be avoided. Fence off these areas with caution tape. If they can't be avoided, plan for a pump.

■ **Fences and stone walls.** Chain-link fences and wooden fences can be temporarily moved without too much effort, but moving a stone wall would be more costly than bringing in a pump truck.

■ **Property lines.** Concrete trucks can leave deep ruts in a lawn, so be alert to property lines. Laying down wide planks or sheets of plywood under the wheels may help reduce the damage.

■ **Underground sprinklers.** Flag the head locations so they can be avoided. They are not costly to repair, but it might be worth considering the expense of a pump compared with replacing sections of crushed pipe and broken heads.

■ **Landscaping.** Include the cost of avoiding and/or repairing gardens and other landscaping.

Wet weather exacerbated the damage heavy machinery caused to this lawn.

and temporarily fastening the new forms to them is not a problem.

MODIFIED PIER FOUNDATIONS. These are old pier foundations that over the years have been modified by filling in between the piers with cinder blocks. You never really know what to expect with these. The filled-in section may be dry-stacked blocks, mortared, or even grouted.

The biggest problem with modified pier foundations is that the excavator cannot dig too close to the foundation for fear of causing the earth to move and the foundation to shift and crumble. Removing the last foot or so of earth from against the foundation is usually left for the foundation installer. Also, the base of the pier often is a large, irregular chunk of concrete that was poured into a

dug hole. Matching forms to this irregular edge is difficult and time consuming.

MORTARED STONE FOUNDATIONS. These are normally pretty solid. Marrying the forms to the irregular profile takes a little extra time.

DRY-STACKED STONE FOUNDATION. These can be very difficult to work with. They have the same excavation problems as the modified pier type plus the irregular profile of mortared stone except more exaggerated. Definitely figure in extra labor when dealing with this type of foundation.

HIGHER EXISTING FOUNDATION. Check the elevations of the planned foundation and the elevations of the existing foundation. If the new founda-

A shovel is used to clear earth from against the existing foundation during the excavating.

tion is going to have a lower floor elevation than the existing one, you have to plan for some type of concrete work to stabilize the soil exposed by the different elevations. This typically involves pouring a small single-face wall as explained in detail later in this chapter.

Laying out a bumpout

The layout depends upon how the new addition relates to the existing structure. If the addition is a simple bumpout with walls that are square to the existing foundation but aren't in the same plane, then the steps are easy:

1. MARK THE EXISTING FOUNDATION. Mark the location of the outside walls of the new addition near the top of the existing foundation and draw a plumb line down to the bottom.

2. LOCATE THE OUTSIDE CORNERS. Calculate the diagonal measurement for the foundation as described in "Laying Out Square Corners" on p. 47. Hold a tape against the existing foundation at the bottom of one plumb line, and measure out the distance for that perpendicular wall. Hold the second tape on the other plumb line, and measure out the diagonal distance. Adjust the two tapes until the measurements cross, then pound a stake in at this point to locate one corner. Repeat this process for the other corner.

3. CHECK FOR SQUARE. Check the distance between the two corner stakes. It should equal the length of the new parallel foundation wall. If it doesn't, you probably figured the diagonal wrong.

Bumpout Layout Sequence

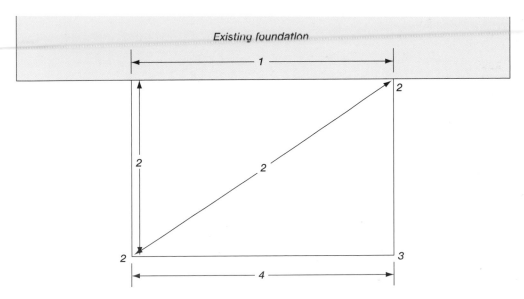

1. Mark the addition on the existing walls.
2. Measure out the proper distance and use the diagonal to locate the first corner.
3. Locate the second corner.
4. Check the length of the outside wall layout.

The
the

Pinning New Work to Old

Where new concrete work meets old, they are typically secured to each other with steel dowels referred to as pins. Drill ½-in.-dia. holes at least 6 in. deep into the old work. Locate the holes about 1 ft. apart unless a different spacing is specified on the plans. Roughly center the pins in the wall or footing. Into the holes, pound 18-in.-long pieces of ½-in.-dia. rebar. This will result in a tight friction fit, but for a better bond, drill oversized holes with a diameter of ⅝ in. or ¾ in. and use epoxy filler around the dowels.

Holes are drilled for the rebar pins that will tie the new foundation in with the old.

determine the diagonal, then check the diagonal measurement of the three points. The three points are the corner of the house where the new foundation will meet, the first corner stake along the plane of the house, and the next corner stake parallel to the house. If the diagonal length is not correct, that means that the existing house is not square. If it is only out of square by a fraction of an inch, just go with it; the framing crew can adjust it as they wish when they put down the sill plates.

If the diagonal is off an inch or more, adjust the second corner you located accordingly (see "Square or Parallel" on p. 109). Moving the point that runs along the plane of the house is usually not an option. Plot the points for the rest of the foundation, measuring parallel off the two established planes of the new foundation.

Setting forms for an addition foundation

Once all of the corners are staked out, form and pour the footings as described in "Building Footings" on p. 45. The only difference is that the footings—and the new walls—are "pinned" to the existing structure if possible. Then locate the same corner points on top of the new footing and snap lines from one point to the next to outline the layout of the new foundation walls. With the footings completed, you are ready to set forms for the addition foundation.

1. INSTALL PINS. Locate the inside face of the new foundation. For a continued wall, this means measuring in from the outside corner of the existing mudsill a distance equal to the thickness of the new wall—10 in., for example. For a bumpout, measure in from the plumb lines you made earlier. Drill and pin the

The first panel of the new foundation is held plumb while tacked into position.

areas as described in "Pinning New Work to Old" on facing page. If the existing foundation is loose and has an irregular profile, consider omitting the pins. The vibration of a hammer drill can damage a loose foundation. The concrete will conform around the irregular shape, bonding the walls in much the same way as pins would.

2. INSTALL THE FIRST PANELS.

Insert jump or hook rods into the first panel, and place it against the old foundation with the inside face lined up to the plumb line. Then using a few masonry nails, fasten the panel to the wall by nailing through the side rail into the old foundation.

Set the opposite panel and slide the ends of the jump rods into it to hold it in position. Drops wedges into the jump rods to secure everything in place. Repeat this process at any other location where the new foundation meets the old. Then starting at those first panels, set the rest of the wall as described in "Setting the Forms" on p. 60.

3. RUN STRINGS AND ADJUST THE

FORMS. Run strings along the tops of the forms as described in "Stringing the Walls" on p. 67, and reset the original reference line along the existing foundation and over the new forms. Adjust the forms until the offset difference between the reference line and the string along the tops of the forms along that plane are equidistant. Adjust the forms as needed to mimic the original layout, square or not.

4. ESTABLISH THE GRADE. Open a hole through the framing of the existing house at the location where there will be a passage or opening between the new addition and existing (see the photo below). Shoot the elevation of the top of the finish floor and subtract from that the combined thickness of the new finished floor, underlayment, subfloor, floor joists, and the mudsill(s). Subtract an extra ¼ in. to be on the safe side (it's easier to shim up than to rip down) and use that elevation as the top of the new foundation.

If the future opening between the house and addition is to be very wide or if there will be openings at a couple of places, open holes through the framing to get an accurate representation of the level of the existing floor. Sometimes the house is out of level. If this is the case, you must decide whether to match just one spot or build the new addition out of level to match the old. Once the decision has been made, shoot and mark the new grade accordingly.

5. KICK THE FORMS. Short end walls can get pushed away from the existing foundation by the force of the concrete as it is poured into the forms. Bracing the forms, known as kicking, keeps them in place.

To establish the grade of the new foundation, a hole is made to expose the existing finished floor height at the point where the existing house will be opened to the new addition.

Tack an upright piece of 2-by stock on the outside of the form, about a foot in from the corner. Is should be tall enough to extend the entire height of the panel. Lean another upright directly opposite it against the bank. Wide stock such as a 2×8 or 2×10 works best against the bank. Measure the distance between the two uprights, starting near the bottom and at about every 2 ft. going up. Add about 1 in. to the measurements and cut 2×4 stock to length.

Starting with the lowest, tap the 2×4 brace in between the uprights until it fits tightly into place. Work your way up, firmly tapping the rest of the braces in place. You may have to retighten the lower ones if the upright against the bank shifts slightly. Tack the braces into the

Bracing the forms hard up against the bank, referred to as kicking, keeps short addition walls from being pushed away from the existing foundation as the concrete is poured.

When there is no bank to kick up against, digging a hole and lining it with a bucket serves as a good base.

uprights with a couple of duplex spikes on each end.

On sites where there is no bank to brace against, dig a hole at an angle about 8 ft. away from the forms into solid ground. Line the bottom and back of the hole with wide 2-by stock (sometimes a 5-gallon bucket or even some large, flat rocks will do the trick). Have the ends of all braces terminate in the hole. Tack the center of the long braces together for extra support (see the photo at bottom p. 113).

6. FILL IN THE VOIDS. If the existing foundation has an irregular profile, there will be voids between it and the end forms. These voids must be filled to keep the concrete from oozing out. However, as long as the slump is kept low, voids less than 1 in. wide will be plugged by the concrete itself.

It doesn't take much to keep stiff concrete from working its way out of the voids. Small rocks can be tapped into the holes as well as just about anything else, even scrap wood or disposable coffee cups. Pieces of ridged Styrofoam work best because they compress to fit in a smaller opening and expand a little to conform to the area and hold themselves in place. Take more care filling in near the bottom of the forms; the lower you go, the more pressure is exerted by the concrete.

For very large voids such as those found against dry-stacked stone foundations, it is better to scribe lengths of ¾-in. plywood to fit the profile. Make the piece wide enough to extend into the forms about a foot and nail them to the inside face of the panels, one on each side. Tie the free ends of the plywood together with metal strapping about every vertical foot.

Spaces between the existing foundation and the adjacent panels are plugged to keep the concrete from oozing out.

Ordering and pouring

Order a low slump of about a 3 or 4 for the job. Once a few critical areas are hit, you can have the driver add water to loosen it up a bit.

Start pouring near the outside corners that you kicked. Pour until the corners are filled about halfway. This weight helps the braces anchor the wall to prevent shifting. Next, slowly pour a few feet away from the existing foundation, guiding the new concrete against the old wall (see the photo at left on facing page).

Concrete is poured away from the existing foundation, then gently brought up against it by hand.

Keep an eye on the filled voids (if any) on either side of the wall. If you see the material used to fill voids move, stop the pour and move on to the next section. Fill one end about halfway up, and repeat the process at any other point where the old and new meet. If there are any drops in the new foundation, fill those next (see "Installing Drops" on p. 80). Fill them until the concrete is a few inches above the bottom of the drop.

Move back to the first end near the existing foundation, fill that within a foot or so of grade, and then do the same

A Simple Way to Detect Form Shifting

The forms never fit tightly against an existing foundation. So just before you pour, lay a spike on the top of the forms with the head up against the existing house. If a small gap appears between the spike and the house during the pour, you know the forms are shifting away from the house.

In a view looking down from the top, a nail is placed on top of a form up against an existing foundation. If a gap begins to form during the pour, that means the forms are not sufficiently kicked and the forms are moving away from the foundation, increasing the length of the wall.

to the others. Pouring the ends in lifts like this gives the concrete time to set up a bit. This exerts much less pressure than pouring up to grade in one shot. Now you can have the driver mix more water into the concrete until it's about a 5 to 6 slump. Finish pouring and grading as detailed in "Pouring Techniques" on p.84.

Pumping Concrete

When access for the concrete trucks is a problem, the only way to efficiently get the concrete into the forms is by using a pump. There are basically two kinds of pumping systems: a boom pump that carries the hose up and over along a series of booms, and a line pump that has a hose that lies on the ground.

Boom Pumps

These are very large trucks that, in addition to a hopper and pump, contain a series of interconnected hydraulic booms with 4-in. pipes attached to deliver the concrete. At the end of the last boom, a length of 4-in. flexible hose allows the concrete to be precisely directed into the forms.

Make sure there is a fairly level spot for the boom truck with enough clear area on both sides to extend the outriggers that stabilize the unit.

There is a considerable amount of force generated by the descent of the concrete on the downside of the boom. This force is augmented by the pounding action of the pump. This pounding force causes a lot of extra stress on the panels and hardware as the concrete is blasted into the forms. It can also cause segregation of the concrete.

To slow the speed of the concrete caused by the descent, two 90-degree elbows are fastened to the end of the flexible hose, called a gooseneck. Some pump trucks stock two 180-degree elbows fastened together. Also, if stocked by the pump truck, a 3-in. reducer and short length of hose is attached to the end of the elbows for easier placement.

To help reduce stress on the forms and prevent segregation, start the pour by lowering the hose a few feet from the bottom of the form instead of letting the concrete fall 8 ft. Raise the hose as the form fills. Once the concrete is within a few feet of the top, you can work your way around without having to insert the hose back down into the forms.

A boom pump can reach jobs that are inaccessible to concrete trucks. Often it pumps up and over a two-story house to reach a backyard addition.

A line pump has a smaller-diameter hose than a boom pump so it can be maneuvered in smaller areas. However, the large aggregate is limited to about ⅜ in. or smaller.

The pouring procedure is the same as described in "Planning Strategy" on p. 83. Because the pump is so powerful, the concrete can be pumped at a low slump where needed. But the movement of the concrete through the system accelerates the setting time, so there is less time to finish off the concrete. The rate of flow is controlled by the pump/boom operator as directed by the person in charge of the pour.

Line Pumps

These are smaller units that are either mounted on mid-sized trucks or towed on trailers. A flexible 2-in. hose supplies the concrete at a very slow rate. It takes about 45 minutes to pump 10 yards. Because the pump is smaller and less powerful than the boom truck, a special mix of concrete that contains only a small amount of large aggregate (which in this case is only about ¼ in. to ⅜ in. big) must be used. The mix is about $10 to $15 more a yard than regular concrete.

You should start the pour at the farthest point and work your way back, disconnecting sections of hose as you go. The mix is not very stiff, so if there are any drops in walls you may have to wait longer until the concrete sets enough to fill on top of the first lift.

Deciding Which Pump to Use

The minimum charge for a boom pump is twice as much as renting a line pump. But if time is a factor, a boom pump can fill forms as fast as a truck can pour. Also, a boom truck can handle concrete specified with ¾-in. or larger aggregate, which a line pump can't.

The limiting factor for a boom pump is overhead interference. In a congested area, there may be power lines and trees in the way preventing the boom from being used. However, the boom pump can be configured as a line pump, although it is very difficult to handle and support a 4-in. pipe full of concrete.

This addition hole was excavated a couple of feet below the bottom of the existing basement.

A single-face wall is formed and braced against the new foundation, ready to pour.

Single-Face Walls

If the elevation of a new foundation is below the elevation of the existing one, there will be exposed soil that will have to be stabilized. The easiest way to do this is to encapsulate the soil with concrete.

In most cases this can be done with a single-face wall. This is a wall that is formed on only one side. However, if that area of existing foundation is to carry a significant increase of bearing weight (a second story, for instance), consult a structural engineer. He can design a more elaborate supporting structure if needed.

Laying out and forming single-face walls

Exact measurements are not crucial here since the wall's function is simply to keep things in place. Run a string along the length of the exposed bank a few inches up from the cellar bottom. Set the string away from the outermost section of the bank a distance equal to about 8 in. to 10 in. plus the thickness of the form.

Set up forms about ⅛ in. to ¼ in. away from the inside of the string. Lock the forms together with the hardware as you go. Align the tops of the forms with steel channel, or nail long, straight 2×4s to tie the tops together. Run a string over the top of the wall, and align the string to the inside face of the forms at both ends.

Bracing

A considerable amount of effort is needed to brace the wall against the pressure of the concrete. Tack 2-by stock along the length of the forms using one run near the top of the form and another near the bottom. Align the forms to the upper and lower strings. Measure the distance from the 2-by stock to the base of the opposite foundation wall for each panel. Using this measurement, either cut or overlap 2-by stock for braces and wedge them in place between the opposite wall and the 2-by stock on the forms. Nail the braces to the 2-by stock, making sure to keep the forms aligned with the strings.

Braces should be about 2 ft. on center horizontally and 2 ft. to 3 ft. vertically. The wider the space between the forms and the bank is, the more pressure there will be. If there is any question, add more braces.

Pouring a single-face wall

Pour the wall slowly with a 4 to 5 slump mix, being sure to constantly check the form alignment to the string. If the forms start to get pushed out of alignment, stop the pour and tap the braces in a little. Be careful not to adjust the braces too much.

As you get near the top of the pour, there is a lot of pressure on the braces. At this point it's safer to leave the wall out of alignment a little than to try to force the forms in or out.

The new wall formed against the earth keeps the soil undisturbed and in place.

Retaining Walls

While the design of foundation walls takes into account that the top of the wall is supported through the floor system, a retaining wall design includes footings and a network of rebar for support against significant lateral pressure. Most retaining wall specifications (mix, wall thickness, size of steel, etc.) are

Steel in a Retaining Wall

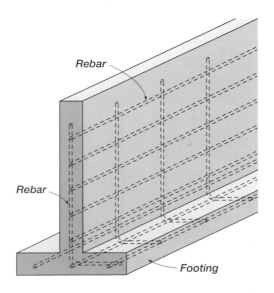

Rebar

Rebar

Footing

■ **WORK SAFE**
■ **WORK SMART**
■ **THINKING AHEAD**

Long 2×10 or 2×12 footing boards make good brace stock for single-face walls. The wide surface area allows for plenty of nailing and pressure distribution.

Striking a foundation tie straight down breaks it off even with the wall.

Flat ties are notched where they meet the wall surface on both sides of the wall. These notches provide a starting point for the metal to break. Using a hammer, strike the ties straight down on their edges. The tie will break off cleanly.

The ties used in the corners of the forms are more deeply embedded in the walls, so you can't take advantage of the notches. The easiest way to break these is to use a reciprocating saw with a metal blade or a hacksaw to cut about ⅛ in. into the top of the tie. Then use a hammer to snap it off.

REMOVING FOUNDATION RODS.

Foundation rods can be a little more difficult depending on the type of rod and the strength of the steel. Flush with the concrete, there is a flat section on most rods that is meant to accept the hardware for setting up the forms. Place the claw of a framing hammer over the rod so it locks firmly into the flat area, then hold the head with one hand and twist the handle of the hammer. If the rod does not snap off, then you have to use the next method.

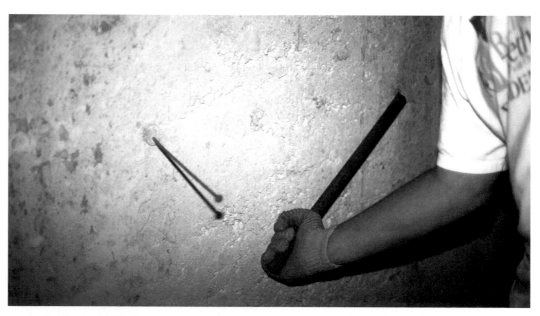

To snap off a foundation rod, slip a pipe over the rod, bend it toward the wall, then rotate it in a clocklike motion.

Using a ½-in. steel or copper pipe about 1 ft. long, slip it over the rod all the way down to the wall. Bend the rod until the pipe is close to knuckle-scraping distance off the wall. As you move the pipe in a circular motion like the hands of a clock, the rod should snap clean at about 180 degrees into the turn.

If the rod is the round, straight type as opposed to deformed or flat-profiled, then the whole rod will rotate in the wall instead of breaking off. In this case, lift the pipe away from the wall and bend it toward the wall in the opposite direction. A couple of bends like this will fatigue the metal, causing the rod to break off flush with the wall.

SHAVING THE FINS. During some pours, concrete oozes out from between panels. This concrete cures into thin "fins." The more the concrete was watered down during the pour and the older the panels, the larger and more frequent the fins. A flat-bladed shovel, a garden spade, or even a T connector works well to break off the fins. One quick pass along the fin will shear it right off.

Sweeping the Walls

Just before you apply sealant, use a stiff broom to sweep the outside of the foundation walls clean of any dirt that may have splashed on during rainstorms or been blown on by the wind in dry weather.

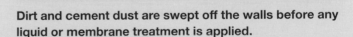

Dirt and cement dust are swept off the walls before any liquid or membrane treatment is applied.

Small amounts of concrete that squeeze out between the form panels create fins that should be scraped off the wall. Here a T connector is used as a scraper.

The inside corner between the wall and footing is eased with a cover of mastic before a sheet membrane is applied.

The backing is peeled off the waterproofing sheet membrane as it is pressed onto the foundation walls.

forced into the corner by the backfill. Also, the area that is to be covered with the membrane must first be primed with a sealer so the membrane will adhere properly.

The installation instructions can vary a little between manufacturers, but most brands are applied in pretty much the same way. A crew of two works best. Once the preparation is completed, precut lengths of membrane long enough to end on the vertical surface of the footing. Draw a plumb line on the foundation and align the first sheet with it. Peel back a foot or so of the release paper at the top and have one crew member press and hold it in place while the other works his way down the sheet, peeling off the release paper and pressing the membrane to the wall.

Continue applying the sheets in this manner, overlapping the previous sheet the proper amount according to the manufacturer's instructions. (Most brands are printed with lines for the proper amount of overlap.) A small hand roller like you would use for plastic-laminate work is helpful to ensure that the overlaps are well sealed. Follow the instructions for the proper treatment of joints and termination edges.

CEMENTITIOUS MEMBRANES. These are often applied to the inside surface of a foundation to try to rectify an existing water problem caused by a poorly treated exterior surface. Occasionally, this works. The biggest problem with these membranes is that they have very little, if any, ability to stretch across any cracks that form after the membrane is applied.

These membranes usually come as a powder that you mix just before application. A liquid acrylic additive is used in place of water or in addition to water to give the membrane better bonding capability and durability. Once you mix it to

Snapping Grade Lines

The last thing you want to see above grade on a foundation is the damp proofing or waterproofing. The damp proofing or waterproofing should terminate 2 in. to 4 in. below the finished grade. Work with the excavator to determine exactly where the finished grade will be at points all around the foundation. If the wall treatment is exposed, it is typically near any steps in the foundation. Snap chalklines or use a lumber crayon to mark the top of the wall treatment.

Lines representing the varying grade heights are snapped onto the foundation as a guide for the wall treatments.

the right paintlike consistency, brush it onto the surface.

Providing perimeter drainage

The second part of an effective surface-water plan is to install perimeter drains to carry the water away from the base of the foundation. The exact details of this system can vary depending on the region, soil type, topography, and available materials, but it is important to have some sort of base drainage system in place to relieve the hydrostatic pressure. Relying on a waterproofing wall membrane alone is not a good bet.

Perimeter drains consist of a continuous perforated pipe that runs around the exterior of the foundation collecting the water as it leaches down through the soil along the walls. Layers of crushed stone and filter fabric are used to wrap the pipe, ensuring a free flow of water and preventing the pipe from eventually getting clogged.

For foundation walls with no footings that sit on a bed of crushed stone, interior perimeter drains are often used instead of exterior. Interior drains work as well because the pipes intercept any water that travels beneath the walls, before it has a chance to work its way up through the slab.

Either rigid perforated white PVC pipe or flexible perforated black corrugated pipe is used for perimeter drains.

The procedure I'll describe uses 4-in. perforated rigid PVC pipe that is available at lumberyards, home centers, and many other suppliers. Often, flexible 4-in. perforated black drainpipe is used instead. It is sold in long coils that are easier to lay down than rigid PVC pipe, and it can be bent around corners so you don't need fittings. It is also cheaper than PVC. The drawback is that flexible pipe crushes more easily than PVC pipe.

1. LAY DOWN GEOTEXTILE FABRIC.

This fabric, also called landscaping fabric, can be found in rolls of different lengths and widths at foundation-product suppliers and landscaping suppliers. Start with a piece of fabric that is about 6 ft. wide. Cut it long enough to extend past the walls by a few feet on both ends. Lay the fabric on top of the footing, and extend it down to the ground and out from the wall.

2. PLACE CRUSHED STONE. Place

6 in. to 8 in. of ¾-in. crushed stone on top of the fabric up against the footing and extending about a foot from the footing. Use a rake or flat shovel to level the stone.

3. INSTALL THE PIPE. Use 4-in. per-

forated PVC pipe to form a continuous drain around the perimeter of the foundation a few inches away from the footing. If there are a number of small, closely spaced jogs in the foundation, it's not critical to exactly follow the foundation perimeter with the pipe. As long as the pipe sits firmly on the stone bed, just get it close. Be sure to face the holes of the pipe down. Gluing the pipe together is not necessary; the friction fit between pieces and couplings is enough to hold it in place as long as the next layer of stone is put on gently.

Perimeter Drain

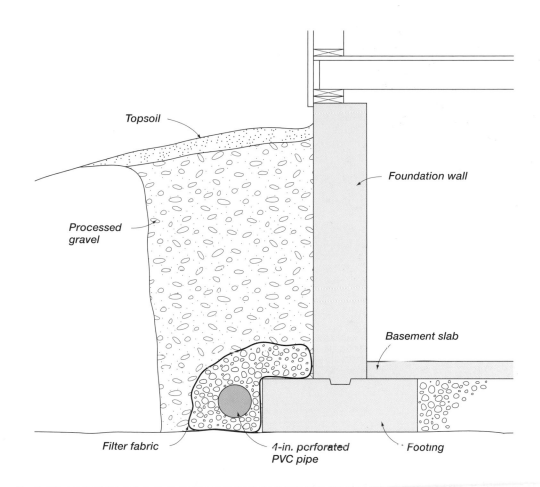

Topsoil

Foundation wall

Processed
gravel

Basement slab

Filter fabric

4-in. perforated
PVC pipe

Footing

Some installers prefer to pitch the pipe slightly toward the outlet. This is not necessary since the holes are on the bottom of the pipe. And unless there is also a problem with rising ground-water, most of the water will bypass the pipe and drain through the bed of crushed stone.

4. COVER THE PIPE WITH STONE.
Place another layer of crushed stone on top of the pipe and the footing until it is a few inches above the bottom of the wall. Pull the filter fabric up and over the crushed stone, and hold it in place against the wall with shovelsful of stone.

Discharging the perimeter drain

You must provide a way to discharge the water from the perimeter drain. The ideal way is a gravity system in which you run a discharge pipe from the drain out to daylight, making sure it's pitched down. This of course means that your lot must slope down enough so that the end of the discharge pipe will eventually emerge out of the ground.

If you have the slope for a gravity system, have a trench dug from the foundation hole to the point of discharge. Continue the filter fabric and the 6-in. to 8-in. bed of crushed stone from the drain to about 10 ft. away within the trench. Put a T connector in the drain, and con-

Perimeter drains are installed on the inside perimeter of the foundation and connected to an exterior pipe (green on the far left in the photo), which will pitch away from the house and drain to daylight.

nect a length of perforated pipe to the T heading into the trench. Cover the pipe with about a foot of crushed stone, and wrap the filter fabric around it. Attach lengths of solid pipe to the discharge until it protrudes out of the finish grade. Cover the exposed end of the discharge pipe with screening to keep the wildlife out.

DISCHARGING TO A DRY WELL. If your lot does not slope enough for a gravity system, then there are a couple of other options. You can have a dry well installed away from the foundation that will accept the discharge and disperse the water. This involves some sort of concrete container, such as well tiles or floater fuses, which are hollow 4-ft. by 4-ft. concrete blocks cast into a waffle pattern. The container is filled or surrounded with crushed stone and gravel. An experienced excavator familiar with the surrounding conditions

Dry Well

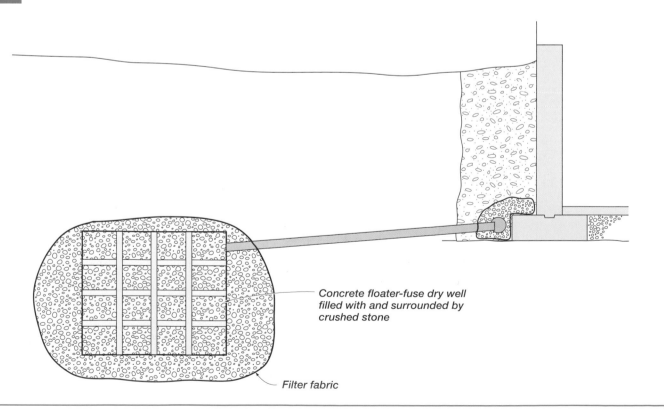

Concrete floater-fuse dry well filled with and surrounded by crushed stone

Filter fabric

can usually recommend a design, but to ensure proper sizing, the system should be designed by an engineer.

The biggest problem with a dry well system is that if you have a high water table, during a particularly wet season the water may rise enough to submerge the dry well. If this happens, the perimeter drains will have nowhere to discharge. This is another reason to have an engineer design the system because he will take the water-table level into account.

DISCHARGING TO A SUMP PUMP.

The other option is to run the discharge pipe into a sump pump located inside the foundation. Some builders frown on this method, reasoning that you should never purposely bring water inside the foundation. Others routinely include this practice in their systems with success.

If you choose the sump-pump method, plan to run a sleeve through the

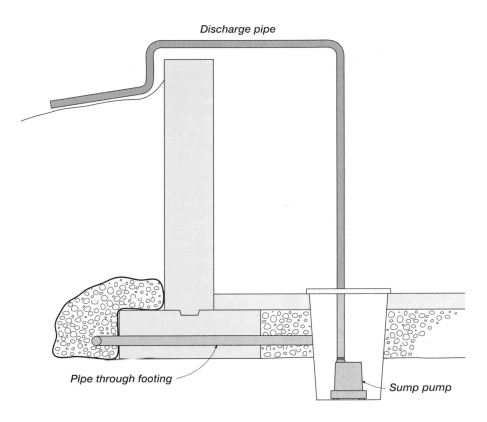

Pipe to Interior Sump

Discharge pipe

Pipe through footing

Sump pump

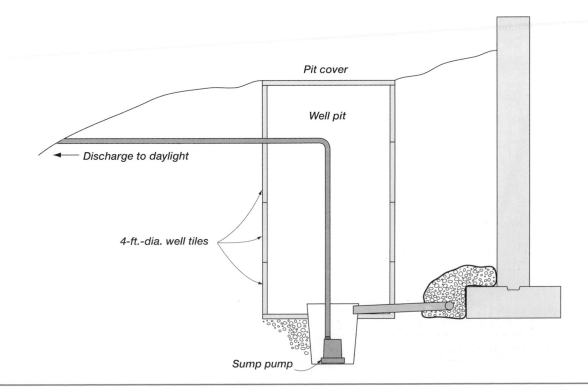

Exterior Sump and Pit

Pit cover

Well pit

Discharge to daylight

4-ft.-dia. well tiles

Sump pump

footing big enough to slip the 4-in. discharge pipe through. This eliminates the need to dig beneath the footing, thus undermining it. As with all of the other elements of the house, the location of the drain, discharge, and sump pump should be planned for in advance. An alternative to this is to install an exterior sump pump as detailed in the illustration at the bottom of p. 137. However, because of the excavation and the need for a crushed-stone floor and some sort of wall system such as concrete well tiles with a cap, this can involve a considerable extra expense.

Backfilling

Backfill on top of the perimeter drain with a sandy soil or soil with gravel mixed in. This lets the water percolate through into the drain system without clogging the filter fabric. If you backfill on top of the filter fabric with soil containing clay or silty material, the filter will be sealed within a relatively short time, rendering the system almost use-

less. Ideally, you should backfill the overdig with gravel up to within about 1 ft. of finish grade. This keeps the surface water draining continuously down the wall and into the drain system.

The finish grade around the foundation must slope down and away from the foundation. This carries away a significant amount of water that otherwise would find its way down alongside the foundation walls, greatly increasing the hydrostatic pressure and potential for leaks.

Groundwater

Groundwater is the naturally occurring water below the earth's surface. In most places the level rises up and down with the seasons. Occasionally there is a rise in the water table that is well above the norm. This might happen once in a period of 20 or 30 years or maybe even only once in a lifetime. Regular homeowner's insurance doesn't cover water damage caused by this type of occurrence. So if the basement floor is within a few feet of the water table, it is prudent to plan for this occurrence, especially if you plan to finish the basement.

Sometimes surface water can collect beneath and around the foundation because of geological characteristics that happen to form a kind of bowl below grade. In this case, the surface water is treated as a groundwater problem.

Laying a crushed-stone bed

Lay a 6-in. to 8-in. bed of ¾-in. or larger crushed stone under the entire basement slab. This allows the water to move freely toward a discharge system, reducing the hydrostatic pressure buildup and the potential for leaks. Installing crushed stone beneath a slab also acts as a capil-

Sloped Finish Grade

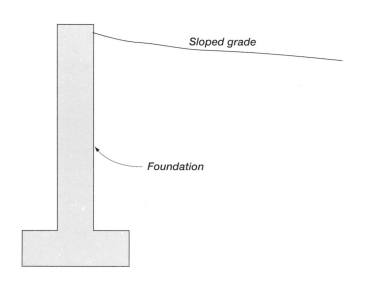

Sloped grade

Foundation

lary break. This prevents water from wicking up into the slab and causing moisture problems in the basement.

After the footing is poured, fill the interior of the footing with crushed stone up to the top of the footing. Some designs call for a 12-in.-wide wall sitting on the ground with no footing. In this case, lay 6 in. to 8 in. of crushed stone in the bottom of the excavated hole before the forms are set. Spread the stone about a foot or so past the perimeter of the foundation layout (see the photo at right).

Installing a drainage system

The crushed stone allows the water to move freely, but a drain and discharge system must be in place to keep the water level below the slab. As with the perimeter drain, the best discharge system is a pipe draining to daylight. If your lot allows this configuration, run a 10-ft. section of 4-in. perforated PVC pipe inside the basement buried a couple of inches below the stone with the holes facing down. Then continue with solid pipe beneath the footing (or wall if there is no footing) or through a sleeve in the footing provided for the planned drainpipe and out through a trench into daylight.

INSTALLING A SUMP PUMP. If you cannot drain to daylight, install a sump pump in the floor. Locate the sump pump so that its discharge pipe will have the shortest run possible. Dig a hole large enough to contain the unit plus about 6 in. of crushed stone on the bottom and sides. Lay filter fabric in the hole and cover the bottom with the stone. Gauge the elevation so that the top of the unit will be even with the top of the basement slab. Insert the unit into the center of the hole, and fill around it with the stone.

Wrap the top of the unit with plastic, and seal it with tape to protect it during the slab pour. Once the basement floor is

Some foundations, such as this one, don't have footings. In this case, the foundation sits directly on a crushed-stone bed that should extend a couple of feet beyond the foundation for proper drainage.

Pipe to Daylight

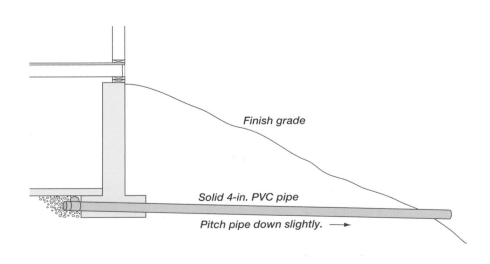

Finish grade

Solid 4-in. PVC pipe

Pitch pipe down slightly. →

Passive Radon Mitigation

As an extra benefit, installing crushed stone makes it easy to mitigate radon. Run a 10-ft. length of 4-in. perforated PVC pipe a couple of inches below the stone with the holes facing down. Attach a 90-degree elbow and a vertical length of solid 4-in. pipe. After the house is framed, have the plumber run the pipe through the walls, into the attic, and up through the roof. He may have to reduce it to a 3-in. pipe to get it buried in the walls. Consult with the plumber when locating the pipe beneath the slab; the vertical section should come up in an unobtrusive spot next to the foundation wall and near the underside of the wall he will be running it in.

The temperature of the pipe rises as it runs from the cool earth through the living space of the house. This warms the air in the pipe, causing a stack effect. The warm air rises, drawing in cooler air from beneath the slab. Radon follows this path of least resistance and flows into the pipe and out through the roof.

Clearly label the pipe as radon mitigation so that some future owner doesn't mistakenly tie in plumbing during a remodel. If for some reason there are still high radon levels, an inline fan can be added to the pipe to change it to an active radon mitigation system.

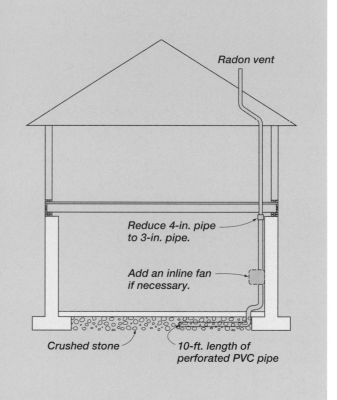

Radon vent

Reduce 4-in. pipe to 3-in. pipe.

Add an inline fan if necessary.

Crushed stone

10-ft. length of perforated PVC pipe

It's a good idea to add a battery backup kit to a standard sump pump. The kit includes a battery, charging unit, and low-voltage pump.

poured and the house is framed, complete the rest of the installation of the pump's discharge and its power needs. The pump's discharge can be run out into a dry well or to a municipal runoff system if allowed by local authorities. It can also run to an area that naturally drains away from the house, as long as it doesn't affect your neighbors. The goal is to get the water far enough away so that it won't recycle back through the system.

In wet areas where the sump pump is an active part of the system rather than a static unit waiting for that unusual, high water table condition, it's a good idea to have a backup system. There are commercial units that have two pumps available within one tub, but it is more common

Installing rigid foam on the exterior is one effective way to insulate a foundation. In this case, the foam is applied on a foundation that will support a slab.

to install a second unit sharing the same discharge pipe with the first. The cost of a second unit is negligible compared with the amount of damage it may prevent.

With two units you can run the power to each from separate circuits. That way, if a breaker trips on one, the other will still have power. You should also have one of the units on a battery backup. It is not uncommon for a storm to knock out the power when you need it most. There are a few brands available that include a battery backup as part of the unit.

Insulation

Creating a warm, dry basement is always a goal. Unfortunately many building-science professionals disagree on standard methods to achieve this result. Some subscribe to the method of installing the insulation on the outside of the concrete, whereas others prefer inside the basement.

Installing insulation against the outside of the foundation walls not only insulates the foundation but also protects

the waterproofing from being damaged by backfilling and later from chafing caused by freeze/thaw cycles. Installing insulation on the interior of the foundation allows the above-grade exterior area to remain exposed to inspect for termites and ants. In fact, in many southern regions where termite infestation is severe, the use of rigid insulation below grade is prohibited. Also, in exterior installations the insulation exposed above grade needs a finish treatment. Insulating the interior eliminates this cost.

Some insulating methods call for plastic vapor barriers, whereas others work better without it. Completely sealing the basement envelope from moisture penetration seems logical, but then when moisture does find its way in there's no easy way for it to leave. To complicate the issue further, there is the matter of the hundreds of gallons of water used to make the concrete. This water needs a place to go as it evaporates out of the fresh concrete walls and slabs within the first two to three years.

Personally, I prefer to wait through at least one wet season following the final grading and landscaping before installing

wall systems or floor treatments (carpet, tile, etc.) in a basement. This allows time for most water or moisture problems that get by your defenses to occur. It also lets much of the water within the concrete to evaporate out. Then I like to install rigid insulation against the interior surface of the foundation wall, strap over it with furring, then drywall.

In choosing a method, be sure to check your local building codes. The insulation requirements for foundations and slabs vary from region to region and code to code.

Installing exterior insulation

There are a few types of rigid panelized insulation used on foundation exteriors, including fiberglass and mineral wool. But the most common type of insulation used with concrete work is 2-in.-thick expanded or extruded polystyrene rigid insulation, such as the Dow® brand Styrofoam. It's available in lumberyards and home centers.

When you use any rigid insulation product, check the specifications carefully. Contact with certain solvents (as in waterproofing and damp proofing) can destroy certain rigid insulation products. For most residential applications, rigid insulation is sold in 2-in. by 2-ft. by 8-ft. pieces with a tongue along one 8-ft. edge and a groove along the other or with a shiplap configuration.

In cold regions, some codes allow the insulation to extend to just below the frost line, whereas others require full coverage all the way down to the bottom of the wall. In either case, the insulation should extend all the way up to the top of the foundation wall to where the wood begins. Most contractors and homeowners keep the top of the insulation to just below the finish grade so it won't be visible, and to avoid extending

the mudsill past the outside of the foundation wall so the insulation butts underneath and is flush to the outside face of the sill. This defeats the purpose of the insulation because it is not covering the part of the foundation that needs it the most. It would be like buying high-performance windows for the house and them leaving them cracked open all of the time.

After the damp proofing or waterproofing is applied, lay a panel vertically at one corner with the tongue extending beyond the corner. You can either precut the lengths of insulation to the required height or leave them long and cut them flush to the top of the foundation later. Use a polyurethane-based glue (the foam type works best) to fasten the panel to the top untreated section of foundation. Try to avoid using nails, especially below grade. Even though the waterproofing should seal around fasteners, there's no sense in puncturing the seal if you don't have to.

If the length of the wall is only a few inches more than an even 2 ft., adjust the first panel to overhang the corner about 6 in. so that the opposite end piece will not be a skinny rip. Continue installing panels along the wall, making sure to interlock the adjacent tongues and grooves (or to overlap the shiplaps). Cut the end panel flush with the corner, and start installing the panels against the next wall. Overlap the panels at the corners to ensure coverage.

Cut around any penetrations and use the polyurethane foam glue to seal any voids. Protect the exposed surface of the insulation by troweling on a stucco or exterior insulated finish system (EIFS) product. Extend the finish at least 6 in. below grade.

Rigid polystyrene insulation is applied to the outside of a foundation.

The void around a sewer pipe through the rigid insulation is filled with expanding foam sealant.

Rigid insulation panels extend 2 in. past the top of the foundation so cap insulation pieces will fit flush.

Interior Rigid Insulation

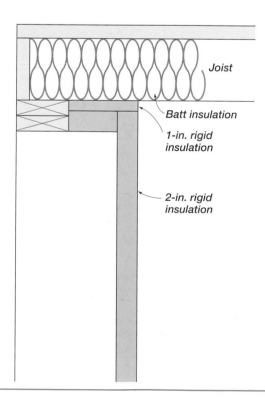

Joist

Batt insulation

1-in. rigid
insulation

2-in. rigid
insulation

Installing interior rigid foam

The downside to using polystyrene insulation on the interior is that, in most building codes, it must be covered for fire protection with ½-in. gypsum board or a similar product. This means you must also install a wood or metal base onto which the gypsum can be secured. Install the panels vertically against the foundation wall, holding them in place with a dab of glue. Ideally the panels should continue all the way up to the bottom of the subfloor by being notched around the floor joists. (First install some kind of insulation in the void between the rigid insulation and the band joist.) But since that height is taller than the 8 ft. the panels are stocked in, the following method is more practical.

1. INSTALL PANELS AGAINST THE WALL. Cut the vertical lengths so the panels extend 2 in. taller than the top of the foundation. Tack the panels against the wall with dabs of foam glue.

2. INSTALL FURRING STRIPS. Snap horizontal lines across the installed panels every 16 in. or 2 ft. on center, whichever you prefer for fastening gypsum board. Place long lengths of 1×3 strapping along the lines, and fasten it through the insulation and into the concrete wall using masonry nails or masonry screws. Place a strip along the bottom of the panels, keeping it ½ in. or so up off the floor. Drop the top strip down a few inches from the top of the foundation to keep the concrete from shattering by nailing too close to the top edge.

3. INSULATE THE TOP OF THE FOUNDATION. Cut rips of insulation to fit tightly between the mudsill and the extended panels. Apply glue on one side of the rips, and wedge them into place so that the glue comes in contact with the top of the foundation. Finish by installing some kind of insulation (usually a loose batt type) against the band joist from the mudsill to the subfloor.

Installing fiberglass batts

Instead of rigid insulation on the interior of the foundation walls, you can use fiberglass batts supported by a stud wall. One advantage to using an interior stud wall for insulating a basement is that you can easily create enough room between the wall and the concrete to accept insulation of any thickness.

First, build a stud wall along the perimeter of the foundation. Locate the

Basement Stud Wall

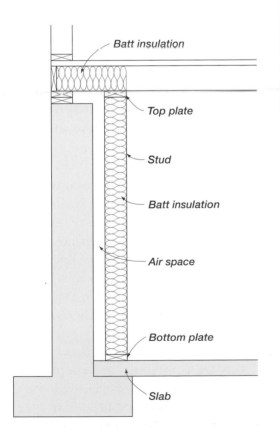

- Batt insulation
- Top plate
- Stud
- Batt insulation
- Air space
- Bottom plate
- Slab

Alternative Ways of Fastening Rigid Foam

One type of rigid panel is rabbeted along the vertical edges to accept strips of wood. The panels are fastened to the concrete wall by nailing or screwing through the wood strip into the concrete. In addition to holding the panel to the wall, the wood provides a 2-ft. on-center nailing base for ½-in. gypsum board.

Another method for fastening rigid insulation is to use metal "Z" channel. Starting in a corner, fasten a Z channel to the wall using a power-actuated fastener. Slip one edge of a section of insulation board into the channel, then hold the other edge against the wall with another Z channel and fasten it to the concrete. Repeat the process by alternately installing insulation board and Z channel. Attach drywall to the Z channel with regular drywall screws.

Studs in Metal Tracks

In place of wooden top and bottom plates, use steel-track plates designed for steel-studded walls. They are less expensive and they isolate the studs from the concrete floor. Also, since the studs bear no load, they don't have to be cut to the exact floor-to-ceiling length. This makes the installation more forgiving on an uneven floor.

Using metal track is an inexpensive way to satisfy the code requirement that wood studs be isolated from concrete.

wall far enough away from the foundation wall so there will be 1 in. of airspace between the insulation and the concrete. The bottom plate of the wall must be pressure-treated lumber since it will be in direct contact with concrete. Fasten the top plate to the bottom of the perpendicular ceiling joists. Install blocking between the mudsill and parallel joists to secure the top plate in that direction. Toe-nail 2-by studs at 16 in. or 24 in. on center. (For more information and techniques about wall framing, see *Precision Framing* from The Taunton Press.) Install insulation along the band joists from the mudsill to the subfloor, extending out to the face of the new stud wall. Next, install appropriately sized insulation between the studs from floor to ceiling.

Even if it is not required, it's a good idea to cover the wall with gypsum board to protect the batt insulation. Make sure to prewire the walls for outlets and switches as required.

An alternative to wood studs is metal studs. They can initially be a little difficult to work with for a carpenter experienced only with wood, but the learning curve is short. The advantage is that metal studs are always straight and they don't bend, warp, or twist, as wood is prone to doing. And of course, termites don't eat metal, always an advantage below grade.

Insulating the slab

Polystyrene is as effective for insulating slabs as it is for insulating walls. Extruded and expanded polystyrene work equally well as long as the expanded is the dense type.

Most typically, the insulation is installed below the slab. After the walls are poured but before the slab is poured, level out the base material (preferably it's crushed stone in a basement). The base should be level with the top of the foot-

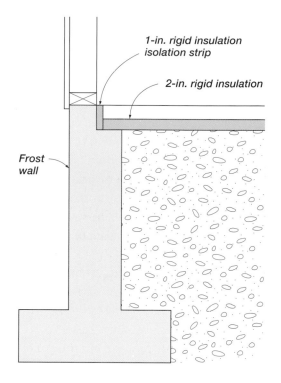

Insulating a Slab on Grade

1-in. rigid insulation isolation strip

2-in. rigid insulation

Frost wall

ings, if you have them. This allows the insulation to extend wall to wall over the top of the footings.

Isolate the slab from the walls with rips from 1-in.-thick panels. Rip the pieces to a width equal to the thickness of the slab plus the thickness of the insulation you'll put under the slab. Typically, this would be a 6-in. rip—4 in. for the slab and 2 in. for the insulation board. Snap a line around the inside perimeter of the foundation to identify the top of the concrete slab. Line the walls with the rips, aligning them with the chalkline, then tack them in place using masonry nails and glue. Next, lay down full rigid insulation panels to cover the entire floor area. Make sure to interlock the tongues and grooves.

■ WORK SAFE
■ WORK SMART
■ THINKING AHEAD

Adding rigid insulation to the floor system raises the floor, giving you less headroom in a full basement. If headroom is important, plan to add an extra mudsill or two to raise the joists, compensating for the lost room.

CHAPTER 6

Flatwork

I N G E N E R A L , flatwork is slab-on-grade concrete work such as basement and garage floors, sidewalks, and driveways. The concrete used in flatwork is usually a stronger mix (about 3,000 psi and up) than that used for foundations, and the aggregate is usually smaller (less than ¾ in.), making the concrete a little easier to handle. Residential slabs commonly are 4 in. thick for normal traffic. To stand up to heavier traffic such as large delivery or utility trucks, driveways are often upgraded to 5 in. to 6 in. thick.

This chapter covers the process of pouring flatwork, from forming to finishing. It's equally important that you be familiar with the concrete basics covered in chapter 2 to ensure that your work cures to a strong and durable product.

Reinforcement and Control Joints

Concrete moves, sometimes a lot. This movement, caused by temperature variations, loading and stressing, and movement of surrounding materials, is to be expected and is not a problem as long as it is planned for. Here I will discuss the different methods used to control move-

A good-sized crew keeps a flatwork project flowing smoothly.

ment and reduce defects such as cracking. Some of the terms and procedures are covered in chapter 2, as they relate to concrete work in general. Here I'll address them as they relate specifically to flatwork.

A badly cracked slab can be avoided by using the correct precautions and procedures.

Preparing the base

Residential flatwork has little structural strength. Unless specifically designed to, flatwork is not intended to bridge over large voids, cantilever, or heave up and down without damage. Proper base preparation, consisting of excavating to stable soil and filling with gravel, is crucial to the longevity of a flatwork project. If the soil beneath flatwork moves or does not provide enough support for the flatwork and its intended load, the concrete will crack and fail.

1. EXCAVATE THE BASE. Before installing a gravel base, remove any topsoil from the area. For exterior slabs, such as patios and driveways, extend the base about a foot beyond the edges of the slab. This gives the slab the uniform support it needs at the edges. At least 8 in. to 10 in. of gravel is required for a good base.

▪ WORK SAFE
▪ WORK SMART
▪ THINKING AHEAD

Install the gutters and diverters before you pour any flatwork along the exterior of the house. This keeps the large volume of water from a rainstorm from washing away the base, undermining the flatwork.

A compactor is run over a gravel base to prevent settling and to provide a uniformly flat surface.

However, soil that contains organic matter or lots of silt or clay may require a deeper sub-base. In these problematic soils, it is not uncommon to excavate out a good 2 ft. or more.

2. INSTALL THE GRAVEL. The top 8 in. to 10 in. of gravel should contain stone no bigger than ¾ in. A sub-base of gravel may be the same material as the top layer, or it may contain larger stone up to about 4 in.

Install the gravel in lifts of about 8 in., compacting every lift using a vibrator. If the gravel is very dry, wet it down first to help it compact. But don't add so much water that you create mud or cause water to puddle up. Of course, you should not wet the gravel in cold weather.

There are two reasons why the compacted base should be as flat and smooth as possible. First, there will be less subgrade drag when the cured slab inevitably moves over the base. Second, the slab should be of uniform thickness so that the entire slab cures and moves in the same manner.

Installing wire mesh

As explained in chapter 2, cracks always form in concrete flatwork. Wire mesh (welded wire fabric, WWF) helps to keep those cracks from spreading open. Some

Making Sure Soil Is Settled

If the flatwork is located over the backfilled overdig area of the foundation, you must make sure that the area is nearly completely settled. Depending on the soil and backfilling conditions, the backfill can easily settle 6 in. to 12 in. or more. If there is any question about whether soil is settled enough to build a slab over, have it tested by a soils engineer. A few dollars spent up front can prevent a failed slab later.

Unless the backfill was manually compacted in lifts, which is not the norm in many regions, it can take weeks and sometimes months of wet weather to settle the soil. Also, if the house was erected quickly, the overdig within the garage walls may never get rained on enough to settle properly.

If the backfill wasn't manually compacted or compacted by rain, you can provide water compaction your-

A misting sprinkler hose is set up along the overdig area to make sure the backfill is completely settled before pouring the apron.

self. Run a lawn sprinkler a few hours every day for a few days so it repeatedly saturates the backfilled areas, then give the soil plenty of time to drain (a few days to a week depending on the soil type and weather) before pouring any flatwork.

Wire mesh to control cracking is placed just before a pour.

builders use it religiously in all of their projects. Others use it only in those applications that are not enclosed by foundation walls, such as patios and driveways. The logic here is that slabs that are enclosed by basement or garage walls can't spread apart enough to substantially open the cracks.

The mesh used in residential work is typically 6-in. squares in sheets of 5 ft. by 8 ft. or in 5-ft. by 50-ft. rolls. Although the 5x8 sheets are a little more difficult to transport and handle, they are much easier to use in a slab because they lie flat. The rolled stock tends to curl back into a roll when it is laid out inside the flatwork forms, creating an added nuisance to an already difficult job.

PLACE THE MESH. To be of any value, wire mesh must be properly placed. Overlap pieces of mesh by at least 6 in. (one row of squares). At control-joint locations (see "Making Control Joints" on p. 152), leave an open seam of several

inches. The mesh should be located about midway in the slab but more often than not, its placement during the pour is neglected and it ends up near or at the bottom of the slab, rendering it virtually useless.

LOCATE THE MESH IN THE MIDDLE OF THE SLAB'S THICKNESS. When using mesh in a slab, many crews lay it on the ground within the forms. As the concrete is poured and raked into place, they reach down into the concrete with the flipside of the rake on which a small metal nub is welded. They use the nub to hook the wire and lift it near the center of the pour with a shaking motion. This is done along the length of the mesh every few feet or so. The problem with this method is that during the haste of a concrete pour, a section can very easily get overlooked or forgotten. Or an inexperienced laborer may pull the wire too close to the surface or not high enough

The wire mesh is lifted onto a small amount of concrete before pouring the slab. This eliminates the need to lift the mesh as the pour progresses.

off the ground, creating an uneven application.

You can place the wire mesh on chairs. But this involves added time, and because you have to avoid stepping on the chairs when you walk on the mesh, it makes it very difficult to move around during the pour.

A third option is to pour and screed the slab to half of its final thickness and then add the mesh. This will ensure a correct, uniform placement, but most crews don't do this because of the extra labor involved.

A simple method that is a sort of hybrid of the other three seems to be the answer for many jobs. Lay out the sheets of wire mesh on the ground before pouring begins. Have the driver run the chute in a zigzag pattern while discharging a small bead of concrete. Lift the mesh up on top of the concrete, and adjust it up or down until it is about at a height equal to half the final slab thickness (2 in. off the ground for a 4-in. slab). Now that the mesh is in place, you don't have to remember to hook and pull it up during the pour and you can easily walk on it as you pour the slab up to level. Contrary to what you might think, walking on the mesh won't make it sink into the wet concrete.

Installing rebar

Steel reinforcing rods are sometimes used in place of wire mesh. In addition to controlling cracks, the rebar adds tensile strength to a project that may be exposed to extra stress such as a driveway that will be used often by heavy vehicles. A typical configuration would be #3 steel (⅜ in. dia.), set on 2-in. chairs in an 18-in.-square grid. The chairs ensure that the steel is placed properly and the 18-in. squares allow plenty of room for walking.

Overlap the steel 12 in. and tie it together with wire. As with the mesh, open seams should be located beneath the planned control joints.

■ WORK SAFE
■ WORK SMART
■ THINKING AHEAD

Don't use bricks as chairs for reinforcement. The bricks will suck up the water from the surrounding concrete, creating a weak spot in the slab and possibly causing it to crack directly above.

▪ WORK SAFE
▪ WORK SMART
▪ THINKING AHEAD

Sometimes mixing and curing conditions are less than ideal. Perhaps too much water was used in the concrete mix or you have reason to believe the slab will be left unprotected in hot weather. Watered-down concrete will shrink more and therefore crack more. Hot weather conditions will cause a fresh, unprotected slab to lose water too quickly and also crack more. Although not a substitute for proper mixing and curing, you might want to add additional control joints in anticipation of more cracks.

Using fiber mesh

Although many builders use fiber mesh in place of wire mesh, this is a mistake. Fiber mesh only helps prevent cracks while the concrete is curing. It does not prevent cracks from widening after the concrete has cured as wire mesh does (see p. 33). Also, fiber mesh does nothing to increase the tensile strength of concrete. Fiber mesh works well for the job it is designed to do, but it should not be used as a substitute for other types of reinforcement.

Making control joints

Since you can't always prevent cracks, control joints provide a straight path for a crack to follow when it does occur. It's like scoring a piece of gypsum wallboard with a utility knife; when pressure is applied you know exactly where the wallboard will crack through. These joints are also known as contraction joints.

A control joint can be sawcut into the cured slab a day or two after it has been poured, or a plastic or metal control joint can be slipped into place during the pour. But the most common method is to create the control joint while finishing the concrete using a tool called a jointer or groover.

Regardless of the type, a control joint should go down one-quarter of the depth of the slab—1 in. deep for a 4-in.-thick slab. Be careful not to go too much deeper than the one-quarter ratio. This could weaken the slab enough so the control joint causes a crack instead of just controlling one.

The location of control joints depends on the size and design of the project. Try to place the control joints so that they divide the slab into squares

The top installation strip attached to a plastic control joint is removed immediately after the joint has been placed.

rather than rectangles. But when rectangles are aesthetically desirable such as in long, narrow walkways, don't make the distance between joints more than one and a half times their length.

Typically, control joints in residential slabs are about 8 ft. to 12 ft. apart for the larger slabs. Smaller, narrow slabs such as sidewalks should follow the rectangle ratio. In addition, control joints should also be placed at every inside corner because they are notorious for cracking.

Making isolation joints

Isolation joints, also known as expansion joints, consist of a flexible material that isolates the slab from other construction members such as foundation walls and footing. This allows each concrete mass to move independently of the other. If they were bonded together, when one moved it would take the other with it, causing cracks.

The inside corners of slabs are prone to cracking. For this reason, the inside corners should receive control joints.

A piece of plastic is used as a slip sheet to isolate the top of the footing from the slab.

⅛" foam sill sealer is used as an insulation joint material. It is less visible than the standard ½"-thick joints.

For joints that isolate the slab vertically, ½-in.-thick asphalt-impregnated fiber is the most common material used. It's available in standard widths, including 4 in. to use with the 4-in. thickness of most residential slabs. The fiber is either glued or nailed in place just before the pour and remains in place. If the slab must be sealed or if the isolation joint needs to be hidden, then once the concrete cures, a strip about ½ in. wide is cut off the top and the void is filled with gray elastic caulk.

To isolate the slab from horizontal surfaces such as the exposed part of a footing, use #15 asphalt-impregnated felt. Thin metal or plastic can also be used as "slip sheets" for horizontal isolation (see the photo at bottom of p. 153).

Pouring Basement Floors

There are a couple of ways to approach pouring basement floors. One way is to pour the basement slab before any framing starts. This gives you great access for the concrete trucks and chutes, enabling the drivers to easily place the concrete just about anywhere you want it. You won't be tempted to add water to the concrete to make it loose enough to push through the network of chutes needed when pouring the job through a couple of small basement windows. This approach also provides a nice work area for the framers and lets them immediately cut and install the basement stairs, which is a convenience for everyone involved.

The alternative is to wait until the roof is on so that weather is not an issue. You can schedule the slab for a specific day and hold to the schedule regardless of the weather. Also, controlling the curing is easier in an enclosed area.

Access for the concrete pour is much easier before the first-floor deck goes on.

Establishing basement-floor grade lines

First, establish the top of the slab by placing a mark 4 in. from the bottom of the foundation wall. Measure down from the top of the wall to that mark and duplicate it at every corner. Measure from the top because it's much more likely to be level than the bottom, especially if the walls were poured without a footing.

Snap chalklines along the walls from corner to corner. You should end up with a continuous line around the inside perimeter of the foundation about 4 in. up from the bottom of the walls. If you happened to pick a high or low spot for the first mark, making the majority of the rest of the line ½ in. or more off the 4-in. thickness, adjust the mark accordingly, duplicate it at all of the corners again, and snap new lines using a chalk of a different color.

Grading a basement floor

If the first-floor deck is framed, you can use it as a reference for the level of the soil base. Cut a piece of light wood stock (1×3 for example) so that when one end is butted up against the underside of the deck or the joists, the other end is exactly 4 in. below the grade line. Use this stick as a gauge to make sure the base is level. Starting from one corner, move in a methodical pattern, checking the level of the base and adding or removing material as needed. Once you have graded the floor, run a compactor over it to smooth out the rough areas and ensure the areas where you added soil are compacted.

If the first-floor deck is not in place, run mason's twine about every 4 ft. to 6 ft. across the top of the foundation and cut the wood gauge to use in a similar manner. Make sure the string is taut. You just have to be careful you're not pushing

■ **WORK SAFE**
■ **WORK SMART**
■ **THINKING** AHEAD

Grade the basement floor as level as possible before the foundation walls go up and definitely before the framing begins. If the floor deck is on and you have to get rid of excess material, you might find yourself trying to shovel it all out through a tiny basement window.

After marking all of the corners from the top of the foundation, grade lines indicating the top of the slab are snapped around the perimeter.

Temperature variations in basement slabs are very slight, so the problems of expansion and contraction are reduced to a point where many residential builders, including me, don't bother with isolation joints.

A wooden gauge cut to the proper length is used to check the slab base using the level floor joists as a reference.

the string up out of position when you are gauging.

Another method is to drive wood stakes into the ground about 4 ft. to 6 ft. apart in rows about 6 ft. to 8 ft. apart. Drive the stakes in so they are about 6 in. above the ground. With two people, position a chalkline so it runs along the sides of a row of stakes, and snap it while holding the two ends even with the grade line on the walls. Next, tack a 6d nail into the stake on the chalkline. Level the grade around each stake to 4 in. below the grade nails, then by eye or by using a straightedge, level the rest of the grade between those spots.

Installing reinforcement and joint material

When the grading is done, it's time to fasten isolation joint material to the wall.

If you're planning to cut out some of the joint material and fill the void with a caulk after the slab has cured, then it's helpful to score or cut through the material about ½ in. down from the top and about halfway through before installing it. This makes it easy for the top ½ in. of material to be pulled off later.

The Vapor Barrier Controversy

A plastic vapor barrier of 4-mil or 6-mil polyethylene will prevent moisture from migrating up through a concrete slab. For slabs that are to receive finished flooring such as carpet, wood, or vinyl, a vapor barrier can be very important to the adherence and performance of the floor. For unfinished basements, the decision whether or not to install a vapor barrier is not so clear. Some building codes require it, some don't. Some builders wouldn't think of omitting it, whereas others never bother with it. The decision is based largely on soil conditions. In areas with soil that drains well and has a low water table, moisture is not a big problem. In soils that don't drain well, it can be a nightmare.

Plastic vapor barriers also help prevent radon from entering through cracks that may occur in the slab.

When it comes to placement of the vapor barrier, even industry experts don't agree. Some say it should be directly beneath the concrete. In other words, the concrete should be poured on top of it. Others say it should be placed with a 3-in. to 4-in. blotter layer of gravel on top. This allows the water in the concrete to both evaporate off the top and drain from the bottom, reducing finishing and potential curing difficulties created by simply pouring on top (see "Plastic Shrinkage Cracks" on p. 38).

If you do pour right on top of plastic, keep in mind that it will take considerably longer before the surface is ready to be worked.

Hold the top of the material even with the grade line, and nail it into the wall every few feet or so with ¾-in. or 1-in. masonry nails. The concrete will hold it permanently once it's poured. If required, install wire or steel reinforcement as described on p.149.

Pouring the concrete

Large basement pours should be divided up into rectangular sections. Plan the width of these rectangles to work with the size of your crew and the length of the screeds.

Set up extra chutes on sawhorses or other varied-height supports to hit the farthest, most difficult-to-reach area first. Work your way back to the easier areas until you can pour directly from the truck, if that is possible at all.

Isolation joints are tacked into place with 1-in. masonry nails just before the pour.

Concrete is poured one section at a time. Once the screeding and floating are complete, the next section is poured.

▪ **WORK SAFE**
▪ **WORK SMART**
▪ **THINKING AHEAD**

In place of grade stakes, you can pour large blobs of concrete in the same spots where the stake would be. Cut 4 in. off the wooden gauge you used to grade the basement, and use it to grade the blobs. Or set up an electric transit with the receiver unit on a grade stick set to indicate the proper level. Then fill in between the blobs to create a continuous strip, keeping it all level.

A check screed is used to level off the concrete just after being placed. A crew member with a smaller concrete rake pulls away the excess or fills in voids as needed.

1. POUR TWO SIDES OF A CORNER.

Starting at a corner, begin placing the concrete near but not against the wall. Pour the concrete in a strip that is about 8 ft. to 10 ft. long and 12 in. wide. Use a concrete rake to push the concrete into position against the wall. Push the concrete up to but not covering the grade line. Using a magnesium trowel, level the strip and smooth it out. Then pour a similar strip along the corner's adjacent wall to form an "L" of concrete.

2. POUR THE THIRD STRIP.

After leveling and forming the second strip, form the third side of a rectangle by pouring a third strip around grade stakes and perpendicular to the other end of the first strip. Push the concrete into position with the rakes, and use the trowel to level and smooth the top. As you level the top around the stakes, pull them out. The three strips of concrete now form a "U" shape.

3. FILL AND SCREED THE RECTANGLE.

Now start pouring concrete against the first strip and between the second and third strips, raking the concrete level by eye as you go. Once a 4-ft. to 6-ft. section has been filled and raked, use a screed to strike the concrete level. The screed is a straight 2×4 or 2×6 that is long enough to bridge the two sides of the section. With one person near each end of the screed, place it lightly on top of the smoothed concrete and, using a short side-to-side sawing motion, pull the screed slowly back in 1-ft.-long pulls. Excess concrete that builds up against the front edge of the screed should be raked back by a third person to make the screeding easier for the other two. If there are low spots that do not fill in, repeat the screeding in the same section after pushing more concrete into the low area.

An alternative to the long two-man screed is a tool called a check screed. It is a wooden or aluminum screed about 6 ft.

to 8 ft. wide mounted on a handle that is handled by one person. It can speed the job by freeing one person from screeding. However, it takes a little experience and a knack to efficiently handle a one-man screed.

4. CONTINUE THE POUR. Once the section has been screeded, continue the pour with another 4-ft. to 6-ft. section and repeat the screeding. If you have extra hands, the pouring and raking can proceed with the screeding process following directly behind.

Floating a basement slab

Float the surface immediately after you screed it. Floating smooths out marks from screeding, pushes the aggregate down, and brings the "cream" (the fines) up to the surface. Floating is usually done with a very large magnesium trowel called a bullfloat. Long sections of aluminum poles are screwed together and attached to the bullfloat.

Homemade Chutes

You can make temporary concrete chutes from a long 2×10 or 2×12 bottom and sides made from 8-in.-wide rips of sheathing. You will have to help the concrete along with shovels, but it will get there. You can make a reusable chute from 12-in.-dia. PVC sewer pipe. Rip the pipe in half along its length and you have two instant chutes.

A length of PVC sewer pipe ripped in half makes a convenient concrete chute.

A section of slab is floated to smooth the surface in preparation for the finish troweling. In the background, a tool with a screen called a jitterbug is used to gently tap the aggregate down from the surface of the concrete. The jitterbug step is not necessary but helps create a smooth surface.

Making and Using Kneeboards

Since you have to start hand-troweling the top of the slab before it is completely solid, you need to use kneeboards to distribute your weight so your knees and feet don't slowly sink into the concrete.

To make a kneeboard, cut an 18-in.-square piece of ½-in. to ¾-in. sheathing. Cut a 2-in. by 2-in. by 18-in. piece of stock, and fasten it to the center of the sheathing extending about 6 in. beyond one edge. This gives you a convenient handle to pick the board up out of the concrete without having to dig your fingers in beneath the boards. You need two boards for every person troweling—one board for your knees to rest on and the other for the toes of your shoes.

Two kneeboards made from scrap lumber provide a place for the knees and toes of the workers when hand-troweling a slab.

You must stop the pouring operation and float the surface before the poured area extends beyond the reach of the bullfloat. Try not to stop in the middle of a rectangle; instead finish pouring one or two complete sections, float, then resume pouring.

As the name of the process implies, the bullfloat should skim along the top of the slab—don't apply pressure; the weight of the bullfloat is all that is needed. Whether you are pushing or pulling the bullfloat, keep the leading edge tilted up. The longer the handle, the more difficult it is to lift it up high enough to raise the leading edge of the float when you are pulling it back toward you. One type of handle allows you to raise or lower either the front or rear edge with a twist of the wrist instead of lifting or lowering the end of the handle.

The sides of the bullfloat may leave little ridges in the concrete. Don't try to smooth out these lines; it's just about impossible at this point, and you will overwork the surface. The lines will be eliminated in the next phase.

Finishing a basement slab

Now you sit and watch the concrete cure. During the floating process and as the concrete sets up, water rises to the surface. This is called bleed water. Before the finish troweling can begin, all of the bleed water must be evaporated. Otherwise, troweling will force the water back down into the surface, causing potential defects (see p. 37). In warm weather the evaporation doesn't take long, but in cold weather it can take hours. The concrete must also be firm enough to support the workers as they finish it.

Once the sheen of the bleed water disappears from the surface, test the first area you poured by stepping on the concrete with one foot. When the indentation left by your boot is ¼ in. deep, the concrete is ready to be finished. On a large slab, you can begin to finish that area even if the entire slab is not ready.

Troweling sometimes requires two hands to smooth out any imperfections. Here, the workers use manufactured aluminum kneeboards that allow easier movement than the homemade variety.

Power-Troweling

Small power trowels look like lawn mowers with huge fan blades. The blades are interchangeable so they can float and finish depending on the blade you use. Power-troweling saves a lot of labor and backache, and the finished product is superior to hand-troweling. But it takes experience to run a power trowel correctly. If you want the glasslike surface only a power trowel can produce but you don't have experience, it's a good idea to hire an experienced operator.

A power trowel is sometimes used to give the slab a desired finish.

Pouring a Rat Slab

A rat slab is a kind of substandard basement slab that is used in crawl spaces. It's only a couple of inches thick, and it is finished by using just a bullfloat. Its purpose is to help keep out moisture, hold the vapor barrier in place, keep out burrowing rodents, and for the fussy homeowner, keep the crawl space clean. It's also a lot nicer for the subs to work on than the dirt.

In some areas, rat slabs are required by the building code; in other areas, it's required by common sense. The sequence for marking, grading, and pouring rat slabs are the same as for basement slabs except the tolerances are relaxed. Once the floating step is complete, your work is done.

A chalkline is held to the perimeter lines snapped earlier, then snapped against grade stakes. Small grade nails are then nailed into each stake at the chalkline to indicate the proper level of the slab.

USE A MAGNESIUM TROWEL. There are two troweling procedures to finish off the surface. The first is using a magnesium trowel. Like floating, the magnesium trowel brings fines up to the surface. It also fine-tunes the level by knocking down remaining high spots and filling in low spots. Start back at the beginning of the pour and work backward, smoothing out the marks left by the kneeboards as you go. The force you need to exert on the trowel is determined by the severity of high and low spots and the extent to which the surface has set up. Occasionally two hands are needed to work the surface as you get near the end.

USE A STEEL TROWEL. The final step is to use a steel trowel. This is what gives the concrete a nice hard, smooth, durable surface. Use the same pattern you used with the magnesium trowel. Again, as you're working toward the last section of slab, you may find it necessary to angle the trowel more and apply more pressure.

Pouring Garage Floors

Pouring a garage slab is almost the same as pouring a basement floor, with a couple of exceptions. Garage slabs are pitched, so the grade lines need to be established accordingly. Also, garage floors usually will be subjected to much wider temperature differences that may make isolation and control joints prudent.

The grade for the base of the slab should already have been determined and snapped on the walls before now (see p. 24). To mark for the top of the slab, just measure 4 in. above the base grade lines and snap new chalklines.

Use the stake method explained on p. 155 to mark the elevations within the garage. If you are using a vapor barrier in the garage, don't worry about the few small holes made by the stakes. Check the level and compaction of the base, especially around the sides where it has been backfilled. Make sure you will be getting an even 4 in. of concrete throughout. Don't let the concrete make up for the low spots in the base.

The foundation typically drops at least 6 in. below the base at all door openings in the garage (see "Garage Foundations" on p. 98). During the back-fill, this dropped part probably got filled in up to grade. Before you begin pouring, dig the soil out off the top of the foundation. At this spot the concrete will flow down on top of the foundation, giving the edge of the slab extra thickness and direct bearing on the foundation.

Forming a garage floor

Unless you will be incorporating an apron into the slab design (see p. 166), you will need to form the front of the garage door openings with 10-ft.-long 2×10 or 2×12 stock. Dig out enough soil along the outside of the opening so that the top edge of the form is even with the line for the top of the slab. Fasten the ends of the form into the outside face of the foundation with masonry nails, or drill holes and use masonry screws.

Support the middle of the form by driving a stake into the ground against the form. Drive the stake in far enough so that it is flush with or just below the top of the form. If the soil is hard, drive the stake in until it is nice and solid, then cut the top off even with the form. Backfill against the form to help hold it in place during the pour. You can use the same procedure for entry doors, except that you can eliminate the stake reinforcement.

Instead of attaching a door-opening form directly to a stake, you can pound in the stake about 2 ft. away from the form, then fasten a form aligner, which is a large steel turnbuckle with nailing

The foundation wall is cleared of soil to ensure good bearing at that point for the slab.

Form aligners brace the slab form board. The form can now easily be adjusted in or out as needed after the pour.

The dry gravel in a walkway gets hosed down just in front of the pour to keep the concrete from prematurely drying out.

flanges at each end, between the stake and the form. This will allow you to tweak the form perfectly straight during the pour. Form aligners are available at concrete hardware suppliers.

For double-width garage doors, two or three stakes will be necessary to hold the form in place.

Next, install any isolation-joint material along the inside perimeter of the foundation. At this time also install any vapor barrier and wire mesh.

Pouring and finishing a garage

Just before your pour, wet down the gravel because a very dry base will take too much moisture from the concrete too quickly. But be careful not to make it too wet. Pour the concrete in a rectangle as described for a basement on p. 154. The rectangles are usually the size of the bays.

To keep rainwater from building up against the overhead door, remove ½ in. to ¾ in. of concrete along the outside edge of the slab, the part that butts the forms, and taper it up to the correct slab height at the point where the overhead doors meet the slab. In most cases, this exposed section from the driveway to the door is only about 4 in. deep. When rainwater hits the door and runs down, the severe pitch sheds it quickly away. Of course now the driveway has to be dropped an extra ½ in., but that isn't usually a problem.

An alternative to this is to set the outside form board ¾ in. down from the finish slab grade. Then nail 1×3 stock lengthwise to the bottom of a 2×4 and fasten the 2×4 form to the backside of the overhead door jambs. The bottom edge of the 2×4 should be level with the slab grade, and the bottom of the 1×3 stock should be level with the top of the outside form board.

Pour the concrete behind the 2×4 form to the bottom of the 2×4, covering the 1×3, then pour the concrete in the front of the 2×4 to the bottom of the 1×3. Once the concrete has set enough to be troweled, the 2×4 form board is removed and the edges are finished.

This method creates a ¾-in. lip that the overhead door will settle into and prevents any water that works its way under the door from traveling inside.

1. EDGE THE SLAB. Edging is the process of rounding exposed edges of the slab to prevent them from chipping and to make them safer for foot traffic. It is done with an edger, which is a small, trowel-like tool with a curved edge. Edging, of course, is not necessary for a basement slab where all the edges are contained.

An extra form along the inside of the foundation will form a ¾-in. lip in the slab to keep rainwater out of the garage.

First, run the blade of a steel trowel down at least 1 in. between the slab and the wood forms. Work the blade along the form, breaking the seal between the wood and the concrete. Next, place the edger flat on top of the slab with the curved part between the form and the slab. Run it back and forth along the length of the form, giving the outside edge of the slab a small radius.

Once the edging is complete, proceed with the finish troweling on the slab as described earlier in finishing basements slabs. After completing the trowel work, you may have to go back and touch up with the edger.

2. REMOVE THE FORMS.   Wait a day or so before removing the forms. If you pull them off while the concrete is still green, you run the risk of breaking large chunks off the edges.

Forming and pouring aprons

An apron is a pitched section of concrete slab that extends 2 ft. to 4 ft. from the garage slab to meet the driveway. Aprons are also sometimes used around the perimeter of a foundation to divert the water away from the house. It's not uncommon to see a ½-in. to 1-in.-per-foot pitch, or more, on an apron.

1. LAY OUT THE APRON.   Lay out the corners of the apron with stakes, allowing 1½ in. for the thickness of the forms. Where the apron meets the garage, drive the stakes until their tops are level with the elevation for the top of the slab. Where the apron meets the driveway, drive the stakes down until their tops are at the driveway elevation. Make sure the apron slopes down from the garage. If there is no slope, or worse, if the driveway is higher than the garage slab, then the driveway elevation will have to be lowered. Run string between the outside

An edger creates a small radius on the outside edge of the slab.

pair of stakes, even with their tops, to delineate the outside forms of the apron. About every 3 ft. to 4 ft. along the strings, pound in stakes until their tops meet the string.

2. FORM THE APRON.   Assuming the apron will be 4 in. thick, cut 2×4 stock to form the outside and sloped sides of the apron and fasten it to the stakes. The tops of the forms should be flush with the tops of the stakes. If there is a seam, reinforce it with a scab of 2-by stock on the outside and pound an extra stake behind it for support. As mentioned, when incorporating an apron into the garage-slab design, you don't form the front of the overhead door entries. But you still have to dig out and expose the foundation drops below the door openings.

Wire mesh is placed in the garage and runs continuously out into the apron to tie the two together.

3. TIE THE APRON TO THE SLAB.

Use wire mesh to tie the apron into the garage slab. Lay down sheets of mesh so that they extend a few feet into the garage slab.

4. POUR AND FINISH THE APRON.

Pour and edge the three formed sides of the apron along with the slab. Since the apron is outdoors where it will be exposed to rain, snow, and ice, you should give it a textured finish to make it slip resistant. This is done for most exterior flatwork including patios and walkways.

The most common texture is called a broom finish. There are different types of concrete brooms ranging from steel wire to soft bristle. The impressions left by these range accordingly from coarse to

A stiff-bristle broom is placed then gently pulled along the top of the slab to give it a permanent texture.

fine. After the apron has been floated and troweled, place the broom on the edge of the apron slab and, using a single pulling motion, drag the broom across the fresh concrete to create the desired texture. The broom texture should go perpendicular to the traffic.

In most cases, one pass with the broom should cover the whole apron, but if you need a second pass, overlap the runs as little as possible to maintain a consistent pattern. Avoid making multiple passes over the same area, and shake or rinse the broom after each pass to get rid of any material picked up by the bristles.

Pouring Driveways and Patios

I'll be talking about a driveway in this section, but patios are poured and finished in the same way. The rough grading for a driveway establishes the driveway elevations. If the excavator did the job properly, the driveway base should be relatively flat and it should follow the finished contour of the land. It should also be pitched to prevent puddles. If a patio is against the foundation, it should be pitched away from the house. If the patio is freestanding, it should be pitched in the direction that can best accept the runoff.

Laying out and forming a driveway or patio

Laying out the perimeter of a square patio is very similar to laying out an addition foundation. If one or two sides of

A base of gravel is placed and spread around the driveway before the forms are set up.

the patio continue out in the same plane as one or two house walls, see "Laying Out Continued Foundation Walls" on p. 108. If not, see "Laying Out a Bumpout" on p. 107.

Establish the corners of the slab with stakes as described for aprons on p. 166. Be sure to keep the tops of the stakes level with the top of the driveway slab elevation. Run strings from one corner to the next, and install stakes about every 3 ft. along the string. Cut forms to length, using 2×4s for a 4-in. slab or 2×6s for a 5½-in. or 6-in. slab. Fasten the forms flush with the tops of the stakes.

Using a long straightedge, check and regrade the base if necessary. If the slab is too wide for the straightedge, use a series of grade stakes down the middle.

Next, plan out your control joints and mark the locations on the tops of the forms. Try to divide the slab up into squares with the control joints. For a double-wide driveway, plan a control joint down the middle of its length, then measure out the perpendicular joints accordingly. If a driveway is longer than 40 ft., plan to divide it in half with an isolation joint to give it room to expand against itself.

Using Construction Joints

Construction joints allow you to divide a pour into sections that you can finish before you move on to the next section. This is helpful if you have a small or inexperienced crew working on a large driveway or other slab.

A bonded construction joint is one that connects the sections together to prevent horizontal and vertical movement. Deformed steel dowels are typically used for this.

More common in residential work is an unbonded construction joint that doubles as a control joint. In this case, only the vertical movement is prevented by using either smooth steel dowels between sections or by forming a keyway in the joint. The keyway is also known as a tongue-and-groove construction joint.

There are construction-joint forms manufactured from light-gauge steel that remain in place between two pours. There are also steel keyways that you nail to your own wooden forms. These are removed

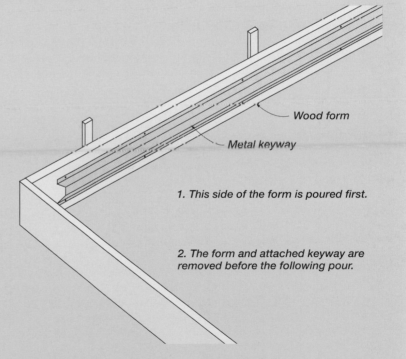

Wood form

Metal keyway

1. This side of the form is poured first.

2. The form and attached keyway are removed before the following pour.

before the next pour is made. Or you can make your own keyway out of wood and fasten it to the wood forms between the pours.

Whichever type you use, the procedure is the same. Form the exterior section of the slab with normal form boards. Use construction-joint forms for any interior sections that will be poured against. Be sure to set any construction joints wherever you plan to locate a control joint.

Front porches, sidewalks, steps, small patios, and landings are all good flat-work projects for a novice to cut his teeth on. Small jobs like these give you plenty of time to do the finish work compared with a larger job where you have to work quickly to get the finish work done before the concrete sets. Plus, the textured finish of exterior flatwork hides small imperfections that would be obvious in smooth interior flatwork such as a garage or base-ment floor.

Pouring and finishing a driveway or patio

Pour the driveway or patio just as you would a garage or basement. Once the slab is ready for hand-troweling, begin by using the edger tool on all edges of the slab that are in contact with forms, except along construction-joint forms (see "Edge the Slab" on p. 166).

Next, snap the guides for the control joints following the marks you laid out earlier on the forms. Place the groover at one end with the blade resting on the line, then place one end of a long straightedge up against the tool. Move the groover about 4 ft. away, over the line, and place that end of the straight-edge against it. Now move the groover back and forth along the straightedge, working the blade down deeper at every pass. When the groover has reached its maximum depth, move the straightedge along the line, gauge it into position with the groover, and continue creating the control joint.

Once all of the edging and joints are completed, continue with the rest of the troweling process. During troweling, the joints may have to be cleaned up a bit.

Give the driveway or patio a textured finish as described for an apron on p.166. Typically, a fine, soft-bristled broom is used for the driveway texture.

Pouring Porches

The front-entry porch described here is enclosed on three sides by house walls. The slab will be supported at the front by a frost wall that is already in place. The frame of the house is complete, and if the siding has already been applied, it has been left out of the porch area.

A worker works on the control joints with a jointer. These joints provide a predetermined path for cracks.

Concrete is often used for the floors of entry porches. Here, the slab was also colored and stamped to give it a different look.

A peel-and-stick membrane is applied to all areas where the concrete slab could come in contact with the framing wood.

Flashing the porch

To prevent the porch slab from coming into contact with wood, use flashing material such as felt paper, roll aluminum flashing, or a peel-and-stick roofing or foundation membrane. Apply the flashing to all wood surfaces that will be in contact with the slab, making sure to extend several inches above the slab.

Establishing grade lines

Establish the top elevation of the porch slab directly beneath the entry door by measuring down from the doorsill to the desired elevation of the slab. Draw a level line at that point extending in both directions to the sides of the porch.

The porch slab must pitch away from the house to prevent water from accumulating and standing against the house frame. Using a long level or a level on a straightedge, transfer the slab elevation you drew under the door to an outside corner and make a mark. Multiply the horizontal depth of the porch in feet by ¼ in. For example, 6 ft. × ¼ in. = 1½ in. Now measure that distance down from the level mark on the outside corner and make a new mark. Snap a line from that new mark back to the first line that runs beneath the door. Duplicate the line on the opposite wall.

Forming the front

The outside edge of the slab should overhang and extend down below the top of the frost wall. This keeps the water running down the face of the slab from entering the seam between the slab and the frost wall. It also looks a lot better. To make this overhang, form the outside edge of the slab as shown in the illustration below.

1. INSTALL THE 2×4 SPACER. Snap a level chalkline across the face of the foundation wall about 1 in. down from the top. Fasten a 2×4 along the front of the foundation with the top edge flush

Forming the Front of a Porch Slab

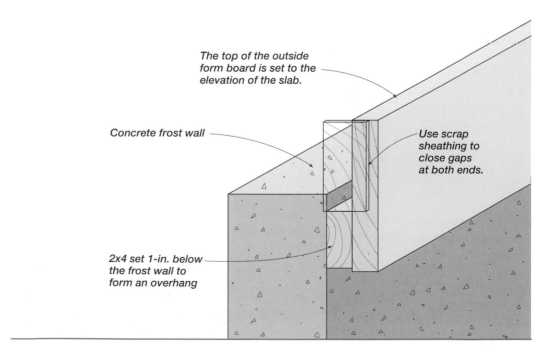

The top of the outside form board is set to the elevation of the slab.

Concrete frost wall

Use scrap sheathing to close gaps at both ends.

2x4 set 1-in. below the frost wall to form an overhang

A form is fastened about
1 in. below the top of the
foundation to prevent an
exposed seam between
the slab and foundation.

with the line. A powder-actuated nailer
firing masonry nails works best.

2. INSTALL THE OUTSIDE FORM
BOARD. Measure from the bottom of
the 2×4 to the grade line of the slab. For
the front form, select 2-by stock whose
width at least equals that measurement.
For example, if the distance from the
bottom of the 2×4 to the top of the slab
measures 9 in., you would use a 2×10 for
the front form. Securely fasten the front
form onto the 2×4 with the top flush
with the grades lines of the slab. If the
porch is long enough to require more
than one length, scab on a scrap 2-by
over the seam on the outside of the form
for reinforcement.

3. CLOSE THE ENDS. Tack on small
pieces of scrap sheathing to close in the
opening between the form board and the
framing at both ends.

The slab form board is set to the top elevation of the slab and screwed into
the lower form board applied earlier.

Holes are drilled and steel rebar pins are pounded in to help support the slab.

Reinforcing the porch slab

In some cases, the house may be framed to leave a lip of foundation around the three enclosed sides of the porch. If so, the slab will be evenly supported on the three sides by those lips and by the frost wall at the front. If there is no foundation lip exposed, then the sides and back of the slab will rest on backfilled soil while the front will be supported by the frost wall.

If there is no lip, anchor the enclosed sides of the slab to the foundation by drilling ½-in. holes into the foundation about 6 in. deep and about 12 in. to 16 in. apart. Locate the holes 2½ in. to 3 in. below the grade line and angle them slightly downward. Pound in ½-in.-dia. rebar dowels about 18 in. long. The dowels should fit snugly into the holes.

Because the holes were drilled at an angle, the dowels will extend above the grade line. Bend the dowels down, following the pitch of the slab so they will

be uniformly buried in the slab. If you want to carry it a step further, use dowels long enough to bridge from the holes to the frost wall foundation.

Pouring and finishing the porch slab

Just after you finish pouring, tap the outside of the forms repeatedly to consolidate the concrete; this will prevent honeycombing along the exposed front edge of the slab. Finish the rest of the job as you would any slab.

Pouring Walkways

Walkways are sidewalks that lead from the front entry and sometimes a side door, usually to a point somewhere up to the driveway. Walkways are easy to pour and finish because their narrow width lets you do all the work from outside the forms. You won't be slopping around the concrete fighting for your footing.

Forming a walkway

To begin, place 2×4s on the ground to form a rough layout of the walkway. Adjust the width and placement of any corners or angles until you're satisfied with the dimensions and pattern. Pound form stakes into the ground wherever the form ends and at the corners, moving them as needed to keep the dimensions true.

Next, establish the elevation at each end of the walkway. The foundation/entry end is determined by the future finished grade of topsoil (with new construction, it may not be in place at this time). The driveway end should match the elevation of the driveway where they meet.

Using the string method described on p. 169, drive in the rest of the stakes and attach the form boards to the stakes.

Normally the form boards will be 2×4s, but sometimes the site isn't entirely graded so you will have to use wider boards to get the correct height.

Touch up the base to make sure it's flat and compacted, and add wire mesh or steel reinforcement. Pour and finish the walkway in the same manner you would a driveway, as described earlier. Make sure you don't exceed the rule of thumb for control joints (see "Making Control Joints" on p. 152).

Building Concrete Steps

Steps are a hybrid of flatwork and foundation. They have exposed flat surfaces for traffic and vertical surfaces for support. The pouring process is similar to that for walls, whereas the finishing work is done like flatwork.

Pouring a footing for steps

A set of steps is a heavy mass of concrete. Unlike other flatwork, steps exert a lot of weight in a relatively small footprint. As with foundation walls, it's best to pour a footing beneath a set of concrete steps to spread the weight out over a larger surface area. The footing also provides a nice flat substrate on which to work.

1. EXCAVATE THE FOOTING. Excavate a hole past the topsoil down to good bearing soil. If the hole ends up being more than 2 ft. deep, you can put 1 ft. of compacted gravel or crushed stone in the bottom or just fill the whole thing with concrete, whichever is cheaper and more convenient.

To support concrete steps in lieu of a footing, some foundation designs include sections of walls poured perpendicular to the foundation walls and about 1 ft. or so lower. These walls are centered under

Forming a Curved Sidewalk Turn

If your sidewalk will make a 90-degree turn, do it with a curve instead of a 90-degree corner. The curve will be more attractive, more natural to walk on, and less likely to crack.

To form the curve of an inside corner, stop the straight forms short of the corner by a foot or so on each leg depending on the radius you want—about 8 in. is the typical minimum. Rip long lengths of ⅛-in. to ¼-in. sheet stock to the same height as the forms (tighter curves will require the thinner stock; ⅛-in. Masonite® works well for a tight radius). Fasten one end of the rip to the inside of the form on one leg. Bend the rip to the desired radius and fasten the other end. The same procedure is used for the outside corners except that the straight 2-by form doesn't necessarily have to stop short. A radius can be formed anywhere within an outside corner. Don't worry about the overlap of the radius form stock over the straight forms because the edging process will smooth the transition.

Radius corners are formed in a walkway by using flexible sheet stock. In this case, ¼-in. plywood is ripped to the correct height and used to form the round corners.

WORK SAFE
WORK SMART
THINKING AHEAD

Plan on forming and pour-
ing the concrete steps
toward the end of the project. Although it may make
entry a nuisance, it will
give the ground as much
time as possible to settle
before being built on.

stair locations and the steps get poured around them.

2. MAKE SURE THE BASE IS COM-PACTED. Steps are normally located over backfilled areas, so make sure the soil in the bottom of the hole has already settled and is well compacted. If not, you can try to compact it with water (see "Making Sure Soil Is Settled" on p. 149), or you may have to continue digging until you hit virgin ground. The sides of the hole should be fairly straight and smooth. If the soil is too loose to accomplish this, use wood to form a rough perimeter footing.

3. DELINEATE THE TOP OF THE FOOT-ING AND INSTALL REBAR. Draw a level line on the foundation at the desired grade of the footing. It should be several inches below the final grade of the topsoil at the location of the outermost step. As leveling guides, pound scraps of rebar about 3 ft. apart inside the hole. Tap the rebar down until the tops are level with the grade line on the foundation. The rebar should be in the ground firmly enough to remain upright as the concrete is placed. If you formed the sides with wood, you can draw level lines and tack in grade nails in place of the rebar stakes.

Footing for Steps

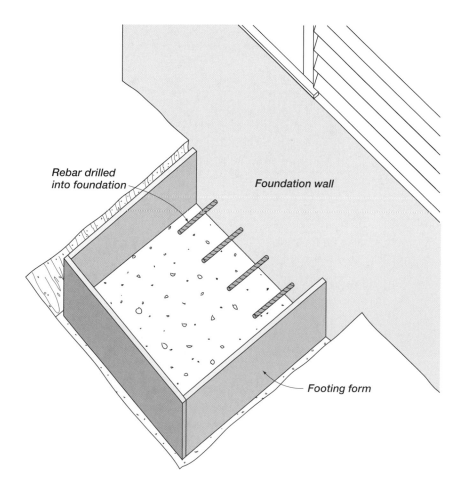

Rebar drilled into foundation

Foundation wall

Footing form

WORK SAFE
WORK SMART
THINKING AHEAD

When using rebar pounded
into the bottom of the footing hole as a depth gauge
for the concrete, put the
orange safety caps on top
of them right after they are
installed. Keep the caps on
until just before the pour.

Calculating the Footing and Steps

The footing should extend about 2 ft. beyond both sides of the steps as well as about 2 ft. beyond the lowest step. To estimate roughly where the lowest step will land (the total run), measure from the doorsill to finish grade. Let's say, for example, that measurement is 65 in. A comfortable step rise is about 7 in., so divide 65 by 7 and round down. This tells you there will be nine risers. The number of treads in any stair is one less than the number of risers, so there will be eight treads.

Since most codes require a 36-in. landing at the top of the stairs, count one of the treads as 36 in. of horizontal run. A comfortable stair tread is about 11 in., so multiply 11 in. by the remaining seven steps to get 77 in. of run. Add 36 in. to 77 in. to get an estimated total run of 113 in. or 9 ft. 5 in. Now, if there is no significant drop in grade, you know the footing should extend about 11½ ft. from the house.

If the grade does drop away from the house, use the method shown in the illustration below to determine how much. Extend a straightedge with a level on top or a line level on a string from the top of the doorsill out to the estimated total run. Measure straight down to finish grade at that point. If the grade drops, add additional steps, one for each 7 in. of drop in elevation. Of course, every step you add extends the footing 11 in. further from the house. If the grade drop is about 4 in. or less, you can take care of it by raising the grade around the lowest step.

To determine the exact unit rise, divide the total rise (65 in., for example) by the number or risers (nine) to get the exact unit rise. In the example, the result is 7.22, which means each riser should be 7¼ in. high.

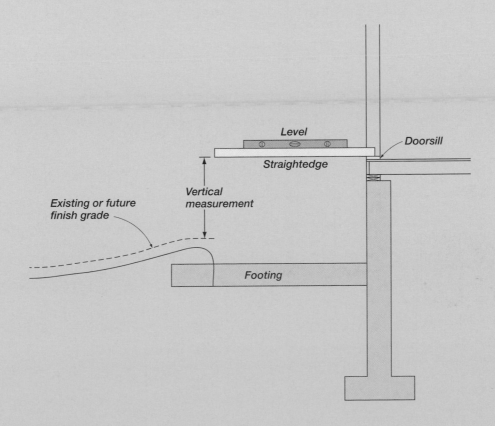

Level

Doorsill

Straightedge

Existing or future finish grade

Vertical measurement

Footing

■ **WORK SAFE**
■ **WORK SMART**
■ **THINKING AHEAD**

For extra reinforcement and security, use metal banding straps to help hold the panels together. Run a few lengths on the footing and a couple behind the panels against the foundation. Once the panels are nailed into position, fold the banding around the 2×4 frame of the panels and fasten it in place.

Drill and install rebar into the foundation as described in "Reinforcing the Porch Slab" on p. 174, being sure to keep the rebar at least 3 in. off the bottom. Place 3-in. chairs or other supports in the bottom. On the chairs, install ½-in. rebar placed in a grid of 18-in. squares and tied together at the intersections.

Pour in the concrete and float it level either with the tops of the rebar stakes or with the grade nails in the forms. If you used forms, remove them after it's cured.

Forming steps

The side forms consist of two pieces of plywood, one for each side, backed with 2×4s. The risers are formed from 2-by stock. Before you install the forms, apply a flashing material such as felt paper, roll aluminum flashing, or a peel-and-stick roofing membrane to all wood surfaces that will be in contact with the steps. Be sure to extend the flashing several inches to both sides of the steps.

1. LOCATE THE LANDING OR TOP-STEP HEIGHT. Measure down from the top of the doorsill one riser height. Make a mark and extend it level across the width of the steps. This line is the top of the steps or landing. Locate the sides of the steps on that line, and draw plumb lines down the foundation.

2. INSTALL REBAR PINS. Drill ½-in. holes for rebar pins to anchor the step to the foundation. Measure from the top of the steps down 10 in. to 12 in., and mark for holes about 6 in. from the vertical lines and about every 1 ft. between. Drill in about 6 in. at a downward angle. Use rebar pins about 18 in. long but for safety's sake, don't pound them in until after the riser forms have been installed.

3. FORM THE SIDES. To form the two sides of the steps, frame two walls with 2×4 studs placed 16 in. on center. Make the walls 1 in. to 2 in. higher than the top-step elevation and a few inches

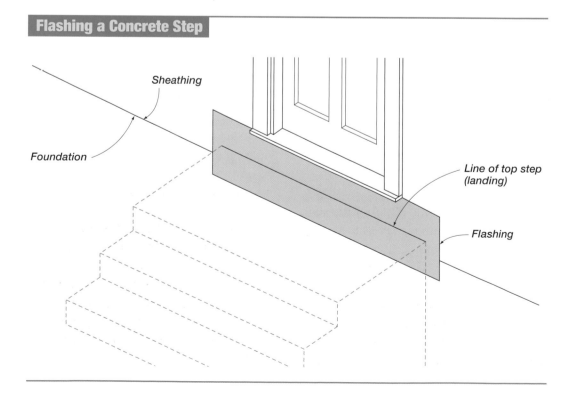

Flashing a Concrete Step

Sheathing

Foundation

Line of top step (landing)

Flashing

Forming Steps

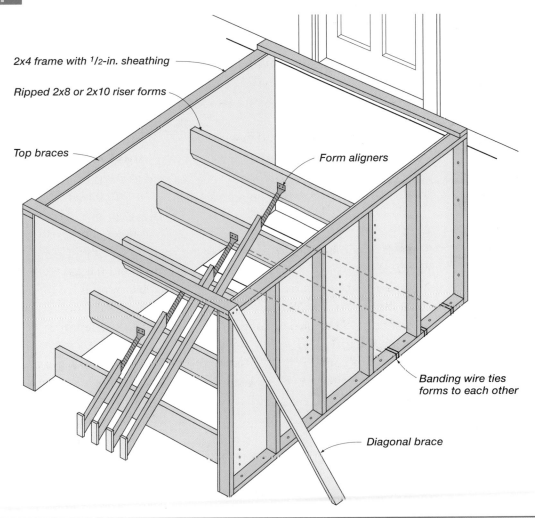

2x4 frame with ¹/₂-in. sheathing

Ripped 2x8 or 2x10 riser forms

Top braces

Form aligners

Banding wire ties forms to each other

Diagonal brace

longer than the total run of the steps. Cover one side of each form with ½-in. sheathing.

4. POSITION AND SECURE THE SIDE FORMS. Place the two side forms in position so that the sheathing lines up vertically with the side plumb lines, studs facing out. Nail the side panels to the house frame and foundation. Move the free ends of the panels the correct distance apart, and measure at the bottom from each inside corner to the diagonal outside corner. Move both ends left or right until the diagonal measurements are the same, then nail the bottom plate of the panels to the footing.

5. CUT RISER FORMS. Cut pieces of 2-by stock to the width of the steps, one for each step, then rip the 2-by stock to the exact height of the risers. Make the rip cuts at a 45-degree angle so that the riser forms won't interfere with the poured top of the treads.

6. LAY OUT THE RISERS AND TREADS ON THE FORMS. Using a level, mark out the risers and treads on the inside face of each panel, starting at the top landing and working your way down. Pitch the landing ¼ in. per foot to keep the water away from the house. Mark the next riser down from the end of

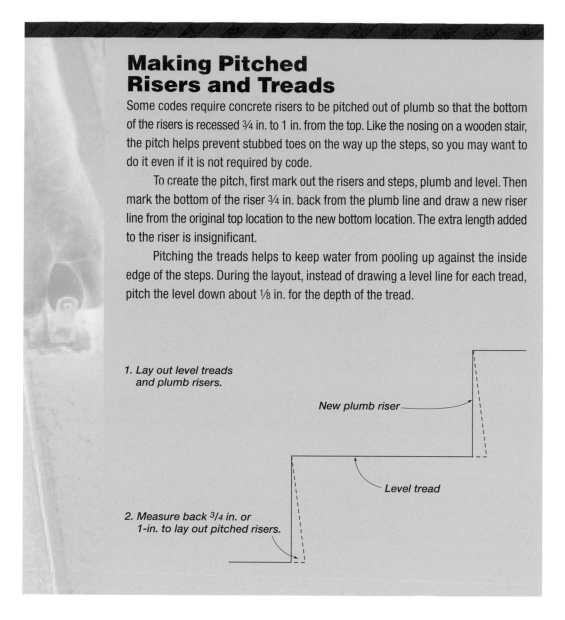

Making Pitched Risers and Treads

Some codes require concrete risers to be pitched out of plumb so that the bottom of the risers is recessed ¾ in. to 1 in. from the top. Like the nosing on a wooden stair, the pitch helps prevent stubbed toes on the way up the steps, so you may want to do it even if it is not required by code.

To create the pitch, first mark out the risers and steps, plumb and level. Then mark the bottom of the riser ¾ in. back from the plumb line and draw a new riser line from the original top location to the new bottom location. The extra length added to the riser is insignificant.

Pitching the treads helps to keep water from pooling up against the inside edge of the steps. During the layout, instead of drawing a level line for each tread, pitch the level down about ⅛ in. for the depth of the tread.

1. Lay out level treads and plumb risers.

New plumb riser

Level tread

2. Measure back ³/₄ in. or 1-in. to lay out pitched risers.

the landing and each consecutive riser from the nose of the preceding one. This keeps the height of the risers consistent.

7. INSTALL THE RISER FORMS. Starting at the top and working your way down and out, insert the riser stock into the panels following your layout so that the flat edge is even with the top of the step, the inside face is even with the riser line, and the bevel is facing out. Fasten them to the forms by nailing through the sheathing.

8. ADD RISER CLEATS. For extra support against lateral pressure, nail 2×4 cleats to the sheathing butting up against the outside face of each riser, especially on the lower risers that will be under more pressure from the concrete.

9. ADD DIAGONAL BRACES. If the width of the stairs is more than 3 ft., split the distance in half and fasten diagonal braces from that point at each riser, down to the ground. Form aligners on the end of each brace make it easy to adjust the risers straight as needed. If the steps are

Formed steps are ready to pour. Note the foundation wall within the steps used for support in place of a footing.

wider than 6 ft., divide the width equally so that no section is greater than 3 ft.

Another way to reinforce the riser stock is to nail a 2×4 on edge lengthwise across the outside center of the board. This can eliminate the need for the diagonal bracing but leaves no way for a quick adjustment if it does bow out during the pour.

10. ADD TOP BRACES AND A CORNER BRACE. Cut two 2×4 pieces to the length of the steps plus 7 in. Nail them securely across the tops of the panels, one at the outside corners and the other just in front of the landing riser. Diagonally brace one of the outside corners to the ground, making sure the forms are plumb.

Pouring and finishing the steps

Start by applying a coating of release oil on the forms. Pour the concrete slowly at about a 4 slump to reduce the pressure and to keep it from overflowing out of the steps. Gently poke it with a shovel or stick as it works its way up the side panels. Pouring from the top of the steps, work the concrete into place at each step with a shovel. Lightly tap the riser forms and the sides of the panels to prevent honeycombing. The more you tap, the more pressure will be exerted against the forms by the concrete, so don't overdo it. Follow the edging and finishing procedures detailed earlier. Let the concrete cure for a few days before removing the forms.

WORK SAFE
WORK SMART
THINKING AHEAD

You can reduce the amount of concrete needed and therefore the amount of pressure put on the panels by filling the cavity with rocks, broken bricks, and cinder block. Pile them against the foundation in a mound and extend it into the cavity, keeping a good 8 in. away from any outside surface.

Index